T0181430

Communicating Genetics

Han Yu

Communicating Genetics

Visualizations and Representations

Han Yu
Kansas State University
Manhattan, KS
USA

ISBN 978-1-349-95460-5   ISBN 978-1-137-58779-4   (eBook)
DOI 10.1057/978-1-137-58779-4

Cover illustration: © Fredrick Kippe / Alamy Stock Photo

Printed on acid-free paper

This Palgrave Macmillan imprint is published by Springer Nature
The registered company is Macmillan Publishers Ltd.
The registered company address is: The Campus, 4 Crinan Street, London, N1 9XW, United Kingdom

Contents

List of Figures

List of Tables

1

Introduction: Visualizing Genetics for Public Readers

It seems banal to repeat that visual representations are an integral part of performing and communicating science; plenty of scholars in science studies, science communication, and science education have convincingly made such arguments (see Lynch and Woolgar 1990; Ford 1993; Kemp 2000; Pauwels 2006; Roth et al. 2005). For the discipline of genetics, visual representations take on an even more significant role. One simple reason for this is the subjects and phenomena under its study are often too minute to directly observe. That observation is afforded when cellular and molecular elements are transformed into inscriptions through the use of specialized visualization tools (Latour 1986). Electron microscopes, for example, give away the shapes of chromosomes, X-ray crystallography reveals DNA's 3D structure, microarray data manifest gene expressions in colorful heat maps, and sequencing technologies turn organisms' genomes into endless scrolls of letters.

For geneticists, genomicists, molecular biologists, and other specialists in fields from genetic engineering, biotechnology, pharmacogenomics, medical genetics, to biochemistry, these tools and inscription procedures would be familiar (certainly, to various degrees). The resultant visual representations, therefore, also come off as familiar and relevant, if not always immediately meaningful and insightful.

© The Author(s) 2017
H. Yu, *Communicating Genetics*,
DOI 10.1057/978-1-137-58779-4_1

1

In the popular communication of genetics, however, the matter becomes considerably muddier. Take the fundamental concept of DNA and the seemingly well-known image of the double helix,

> Most of us have seen images of the double helix, that wonderful spiral strand of "something," first proposed by Watson and Crick. What is its scale? Is DNA larger than a drop of water, a molecule, an atom? What is its color? Does it really look like a spiral? This animated, spiral-shaped image fills the television screen during a report about cloning. It spins clockwise, or is it counterclockwise? It appears in advertising for a biotechnology company wrapped around a typographic logo. It appears in a lively cartoon sequence within the movie *Jurassic Park*. What exactly are we seeing? (Trumbo 2000, p. 381)

As Trumbo implied, the double-helix, at least in the U.S. context, plays multiple roles: It is employed to explain genetics findings and research, it is turned into visual and cinematic entertainment, and it is a physical endorsement of commercialized biotechnology. But can a same image fulfill all of these different roles and purposes? If we cannot even determine its scale, is the image being helpful at illustrating what DNA is and does? Should we, then, set aside the double helix and seek visual representations that portray the structure and function of DNA, like its atoms and bonds? Such images will contain more technical details, but will these details be accessible and relevant to public readers? In addition, such images will probably be less "wonderful" to look at. If so, is that something science communicators should be concerned about when trying to reach non-specialist audiences?

These and other such questions in the popular communication of genetics are what this book ponders. It does so by examining over 100 years of visual representations published in US popular science magazines, high-profile outreach resources, and emerging multimedia formats. The book correlates these visuals to their respective disciplinary, social, and cultural contexts; synthesizes their trajectories in over a century of genetic research and popular communication; and discusses their potential impact on the U.S. publics' uptake of genetics.

Current Research on Science Visualization and the Overlooked Publics

Although science studies traditionally put primary emphasis on the verbal constituents of scientific reasoning and communication, over the years and with attention from multiple disciplines, a rich body of research on science visualization has emerged. These studies examine visual representations that are practiced in diverse fields and disciplines, including anatomy (Kemp 1970; Hildebrand 2004), botany and zoology (Blum 1993; Ford 1993; Pozzer-Ardenghi and Roth 2004), chemistry (Chittleborough and Treagust 2008; Aldahmash and Abraham 2009), paleontology (Gould 1993; Davidson 2008), and physics (Stylianidou et al. 2002; Wickman 2013). A few studies also deal with the use of visual representations in genetics (Amann and Knorr-Cetina 1990; Knorr-Cetina and Amann 1990; Patrick et al. 2005).

Regardless of disciplinary focus, this current literature addresses the perspectives, experiences, and needs of several stakeholders in science visualization. Most notably, we learn about the practices and perspectives of discipline scientists. Studies in this area can be further separated into two groups based on authorship: those authored by social scientists (e.g., Latour and Woolgar 1979; Lynch 1985, 1990; Cambrosio et al. 1993) and those by discipline scientists (e.g., McCormick et al. 1987; Morgan et al. 2006; Krzywinski et al. 2009; Walter et al. 2010).

The social scientists' work, in general, confirms the quintessential role of visual representations in formal scientific practice and provides a sociological examination of that role. Most importantly, the researchers demonstrate that "science" or scientific visuals are not performed and produced in an objective vacuum divorced from social–cultural contexts. Rather, scientific visuals undergo rhetorical maneuvers so as to become publishable evidence (Amann and Knorr-Cetina 1990); they are subject to disciplinary and ideological debates on what is the "proper" way to visualize scientific concepts (Cambrosio et al. 1993); and their "properness" changes with changed historical, disciplinary, and social contexts (Galison 1998).

The discipline scientists' work, on the other hand, describes the everyday reality of science visualization. From this literature, we learn what scientists need in order to best visualize and analyze increasingly complex and high-throughput data: for example, multi-dimensional imaging, visual segmentation of anatomical structures, or integration of multiple visual evidence (Walter et al. 2010). We also learn the latest developments in visualization hardware and software, how these developments overcome current challenges, and how they meet emerging demands.

Besides discipline scientists, science educators and students also had their experiences and perspectives heard (see, e.g., Pintó and Ametller 2002; Stylianidou et al. 2002; Pozzer and Roth 2003; Roth et al. 2005; Dinolfo et al. 2007; Rundgren and Tibell 2009). According to these studies, visual representations are prevalent in science textbooks and play (or should play) a significant role in science education. Unfortunately, as these studies reveal, current images do not live up to that expectation: They have poor data mapping, contain inadequate captions or labels, lack visual cues that guide information processing, and employ surface features that may mislead students (Stylianidou et al. 2002; Roth et al. 2005; Cook et al. 2008). Equally importantly, the literature recognizes that reading scientific visuals is a complex process and factors such as students' prior knowledge and instructor guidance figure into that process (Cook et al. 2008). Science teachers are thus charged with the task to purposefully and effectively guide students' visual learning (Stylianidou et al. 2002; Pozzer and Roth 2003; Chittleborough and Treagust 2008; Cook et al. 2008).

All of these current studies are significant and meaningful in their own ways, but one cannot help but notice that they have largely overlooked one stakeholder group: the publics. Most members of this group do not engage in scientific research as a profession or peruse professional journal publications for the latest experimental results and visual evidence; they also no longer partake in formal education to practice scientific visual learning under the guidance of an instructor. Rather, most US adults, as both national and local studies show, gain information about science and technology through informal materials and channels such as books, magazines, newspapers, TV, radio, the Internet (including online newspapers and magazines), museums, science centers, and interactions with families and friends (National Science Board 2014; Falk et al. 2007).

If visual representations are indispensable to the performing and communication of science in formal research and education contexts, it follows that they should likewise be important (if in different ways) in the informal, popular communication of science. The few studies that turn their attention to such contexts support this conclusion. Northcut (2006), using examples from paleontology, argued that images, though full of dynamic complexities as other aspects of science, are promising at allowing public readers to first access and then assess and participate in scientific arguments. Mellor (2009), citing media representations of genetic modification and particle physics, demonstrated how visuals carry significance beyond their overt meanings to influence public discourse about science.

This book expands this overlooked area of study—and invites future research to do the same. It does so by reporting a larger-scale, longitudinal study on the use of visual representations in the popular communication of genetics, a choice of discipline that, as explained below, is not accidental.

Genetics as a Social (and Scientific) Enterprise

As a discipline that explores the fundamentals of life, genetics is as much a scientific endeavor as it is a social enterprise vested with public interest. Classical genetics from the early twentieth century attracted public attention by offering a way to control the agriculture industry that was vital to the U.S. economy. The eugenics movement in the 1930s, in trying to "regulate" human birth and development, was one of the most publicized (and notorious) scientific initiatives in modern history. As the "new" genetics emerged from the shadow of eugenics to focus on health care and medicine, it became a topic of widespread interest, with mass media daily reporting discoveries of genetics-based maladies and screening, test, and therapy. Also boosting this publicity are debates on controversial topics from Frankenstein food to designer babies, from human clones to genetic discrimination. Probably more tellingly, genetics have been woven into publics' "leisure life" in the forms of fine arts (paintings and sculptures [Anker and Nelkin 2004]) as well as daily entertainments (Sci-Fi movies, TV dramas, and crime investigation reality shows).

And so *should* publics be interested and involved. The impact of genetic research on publics' welfare and quality of life is real and significant—and about to intensify with continuous investigation, government support, commercialization, and international collaboration/competition. As more genetic tests (from prenatal, diagnostic, to predictive) become available and as more diseases (from breast cancer to bipolar) make to the testing list, individuals have to decide whether to subject themselves and their families to those tests and how to cope with test results. When conditions ranging from retinal degeneration to immunodeficiency become treatable via genetic therapies, people need to debate the benefits, cost, and limitations of those treatments. Meanwhile, discussions over DNA fingerprinting, genetically modified food, cloning, stem cell research, and so on require public participation to avoid top-down research policies.

Not only do publics have a vested interest and need to be involved, the field of genetics likewise needs an informed and engaged public. At a time when government research funding is shrinking and public concern over the ethical, health, and environmental risks of biological research rising, scientists need citizens who can relate to their latest findings and sensibly support as well as monitor their work, especially their work translating basic research into biotechnologies and medical/pharmaceutical applications.

For all of these to happen, publics need to be able to access, examine, and evaluate relevant genetics background information as well as latest research findings. "For example, an individual facing a personal decision on gene therapy and an individual seeking to understand the debate over the use of embryonic stem cells in medical research would need to understand the role of DNA, the meaning and functions of stem cells, and the potential use of stem cells in the treatment of selected medical conditions"—among other possible topics (Miller 2004, p. 275). But if we turn to findings from national and international surveys, it is vexing to read that US publics have only minimal understanding on the most basic concepts of genetics. According to the 2000 National Science Board survey, only 29% of the U.S. adults "were able to provide an explanation of DNA that included its role in heredity" (Miller 2004, p. 280). More recent board surveys omitted this question in favor of true/false questions, but the percentage is not likely to have increased, given that US publics' factual knowledge about science has remained static over the past two decades (National Science Board 2014). More recent international surveys confirmed these findings. According to the 2012 BBVA Foundation International Study on Scientific Culture, less than half of US adults knew that ordinary tomatoes contain genes, which put the USA at #10 among the 11 countries surveyed (which are 10 European Union countries plus the USA). According to the same study, only a little over 40% of US adults knew that it is possible to transfer genes from humans to animals, which put the USA at #7 among the 11 countries surveyed.

Certainly, findings from standardized surveys alone cannot reveal what US publics know or do not know about genetics, as these surveys have conceptual flaws. Namely, they are designed with pre-conceived notions of what kind of knowledge matters and is "scientific," and what kind of knowledge should be possessed by "informed" or "cultured" citizens. Almost exclusively, these notions are conceived based on the standards of formal science and divorced from individual citizens' interest, information need, and social and cultural context (Irwin and Michael 2003; Irwin and Wynne 2004; Falk et al. 2007). Relying solely on survey data can therefore result in a deficit view toward the public communication of science. This view, as outlined below, has been thoroughly critiqued and replaced with more fluid approaches that inform this book.

A Heterogeneous Approach to Popular Science Genetics Visualization

The traditional approach to the public understanding of science emphasizes the need for the public[1] to be "properly" educated in what scientists deem relevant and important scientific knowledge (see Michael 2002;

Holliman et al. 2009a, b). To be scientifically literate, one needs to understand specific terms and constructs in physical and biological sciences (for example, the orbiting pattern of the Earth vis-à-vis the Sun) as well as the general nature of scientific inquiry (for example, what counts as a "scientific study") (Miller 2004). It is believed that being scientifically literate will allow individual citizens to properly function in modern society and be motivated to trust and support scientific research. Studies that are grounded in this approach use standardized surveys such as those reported above to reveal the public's inadequate understanding of science and recommend proper education and media communication to reduce that knowledge deficit. In the USA, college-level science courses are advocated, pre-collegiate institutions are called upon to do a better job in science preparation, and science communicators who work in print as well as digital media are charged to "keep Americans up to date about new scientific research and technological developments after the end of formal schooling" (Miller 2004, p. 290).

This deficit view, dominant in the 1970s–1980s and even the 1990s, is now widely recognized as problematic. In addition to relying on standardized surveys that are divorced from publics' individual contexts, it allows scientists to set their own knowledge base and assumption as that which is paramount and thus automatically disempowers the publics to an inferior position. Non-scientists must catch up on their ability and think as the scientists do before they can "properly" participate in conversation or decision making. But as scholars argue, this supposed, clear demarcation between science and non-science, the experts and the "lay" does not really exist, or if it does exist, it is based on problematic social perceptions and should be open to renegotiation (Irwin and Michael 2003). Brian Wynne's (2004) work on Cumbrian hill-sheep farmers is one of the best known studies to support such arguments. In 1986, after the Chernobyl Nuclear Power Plant explosion, thunderstorms washed nuclear fallout onto Britain's upland Cumbrian area and threatened the sheep that grazed the area. In the aftermath, scientists from the Ministry of Agriculture, Fisheries and Food tried to assess and reduce sheep contamination in the area, but in doing so, they ignored farmers' knowledge of the local environment, of hill-sheep characteristics, and of hill-farming management (Wynne 2004). As a result, much of the scientists' formal assessment and experimentation failed.

Studies such as Wynne's (2004) gave rise to a new approach to the public understanding of science, one that is often referred to as the critical approach. Rather than valorizing formal scientific knowledge and emphasizing the public's need to "catch up," the critical approach celebrates citizens' life experiences, local knowledge, value systems, and contextualized information needs. The public thus becomes not one problem group but localized "publics" with varying backgrounds, concerns, and perspectives. And the term "public understanding of science," with the word "understanding" implying deficits, increasingly becomes "public communication of science." With this approach, the cognition of scientific facts is no longer the issue; what becomes important is "the cultural context of the public understanding of science," including the publics' cultural identities, their socially constructed perceptions of science, the effect science and science communication have on those identifies and perceptions, and the dynamic social relationships between publics, scientists, and institutions (Michael 2002, p. 360).

Accordingly, the critical approach rejects one-way dissemination of science from scientists to the publics and instead calls for two-way dialogue and public engagement—although there is no consensus on what exactly constitutes dialogue and engagement or effective dialogue and engagement (Michael 2002; Holliman and Jensen 2009). Some scholars emphasize stimulating publics' curiosity and sense of wonder about science and guiding them to make sense of select topics in science (Meisner and Osborne 2009); others advocate leveraging public stakeholders' knowledge and allowing collaborative decision making between citizens and scientists (Irwin and Michael 2003); still others focus on facilitating direct conversations, Q&As, or debates between publics and scientists via venues such as Café Scientifique, festivals, deliberative meetings, citizen juries, and online consultation (Grand 2009).

While the critical approach mounts a convincing argument against the traditional deficit approach, it is not itself beyond critiques. As Irwin and Michael (2003) admitted, the critical approach can "risk a romanticization of lay people and their relationship to expertise"

(p. 39). The "untutored public" is valorized as being "tutored *in what really matters,* namely, the cultural and the contingent" (Michael 2002, p. 369). Or, in Durodié's (2003) more controversial words, the approach presents public opinions as "values" and thus precludes interrogation of those opinions "by suggesting that we should not offend people's values" (p. 89). In addition, the differences between citizens are ignored and what gets foregrounded is a seemingly harmonious, coherent group that is united in their opposition to the unreflective science (Michael 2002). But in reality, only public members who have strong, preexisting opinions about certain scientific matters will seek to dialogue, while the "silent majority" will remain silent (Thomas 2009). More fundamentally, by emphasizing the publics as "cultural" and implying science as "non-cultural," the critical approach, just as the deficit approach, dichotomizes sciences and humanities, experts and citizens, reinscribing the two-culture split it seeks to denounce (van Dijck 2003).

Trying to rise above these complications is what Condit (1999) called a third option, one that sees the interactions between science and society as more complex, dynamic, and variable. van Dijck (2003) called this the postmodern approach, which foregrounds science as an activity rather than ready-made products waiting to be disseminated to the publics (or disputed by the publics, for that matter). Not only are "publics" inherently multiple (consumers, patients, biotech company employees, etc.), so are the scientific communities, which involve scientists from different disciplines, nations, and cultural backgrounds as well as "non-scientist" experts (e.g., science writers, journalists, artists); given this, rather than the two dichotomous groups of publics and scientists, multiple agents interact to construct and reconstruct science and society in diverse settings from laboratories and journal publications to newspapers, science centers, Hollywood productions, and art galleries (van Dijck 2003).

Other scholars embrace similarly fluid and heterogeneous perspectives, without necessarily calling them "postmodern." Mike Michael (2002), for example, acknowledged that in given settings, either the traditional or the critical approach can be appropriately practical, cultural, and political; at the same time, he recommended a heterogeneous approach that highlights the ongoing, emergent nature of knowledge production and the fluid boundaries between the expert and the lay, the scientific and the popular. Holliman et al. (2009a) likewise emphasized that science communication research can and should be conducted in a number of ways: on one hand, the critical approach facilitates public empowerment, the expression of local knowledge, "two-way" communication, and public debates and decision making; on the other, "an enthusiasm for engagement need not imply an all-consuming abandonment for 'one-way' delivery…. properly fashioned science information can and should continue to be delivered to those eager to know more about the ever-shifting boundaries of frontier (and ready-made) science…and taking full advantage of the enthusiasms and unique understandings that scientists can offer" (Holliman et al. 2009a, p. 276). Indeed, one may argue that in order to realize "two-way" communication and public engagement, some kinds of "delivery" often must exist to provide the relevant scientific context and social exigency (see Stilgoe and Wilsdon 2009; Grand 2009).

These latest perspectives are what the present book draws upon. Doing so, as I explain below, allows the book to approach popular science genetics images as both scientific artifacts and social–cultural artifacts, to acknowledge their diverse functions, and to consider their dynamic impacts.

To start, popular science genetics images—created as they are by scientists or by science communicators in consultation with scientists—function as "one-way" delivery of knowledge, including the fundamentals of genetics, its latest advancements, and its clinical applications. This knowledge, together with other sources of information, allows publics to participate in personal as well as social dialogues and decision making about genetic research and health care. Interpreted by trained scientists, this visual knowledge may be especially welcomed by members of the public who, as Holliman et al. (2009a) wrote, are enthusiastic about learning the "frontier" of science or who, as Priest (2006) put it, embrace the benefits of biotechnology or trust scientists' assessment of biotechnology. For other members of the public, these images may be less eagerly accepted as the ultimate embodiment of "truth," but they still provide relevant information and entry points for discussion.

In order to fulfill these informative functions, popular science genetics images must be made accessible to readers who do not have advanced training or strong prior interest in biochemistry, molecular biology, or genetics/genomics. As scholars elsewhere have argued (MacDonald 2004; Nisbet and Scheufele 2009; Priest 2009), to realize the goal of public participation and engagement in science, popular media need to speak to more than the "elite" and "educated" audience or science "fans" and enthusiasts. Visual representations that are accessible, relatable, and even appealing to a large audience base help to generate wider attention and interest, which is a step, if not the first step, toward dialogue and engagement. In addition, the very process of considering whether and how popular science visuals are grounded in everyday readers' needs and interests helps to avoid exclusive top-down communication and allows two-way or multi-way (re)construction of science.

At the same time as they are delivering formal knowledge, popular science genetics images function as social–cultural artifacts and partake in the public discourse and uptake of genetics. That is, through their aesthetic and stylistic choices, these images embody value judgments about the subject matter they portray and thus influence readers not only cognitively but affectively. In addition, these images are bound up with broader and changing disciplinary and social contexts such as the perceived legitimacy of genetics, the debate over genetic engineering, the euphoria surrounding genetic medicine, the reality of popular science publishing, et cetera. The creation and interpretation of these images is therefore never a purely scientific process, nor a static one.

It needs to be noted that in examining popular science genetics images as social–cultural artifacts, the book does not assume that these images, or media coverage in general, have a linear influence on publics' opinion and perception. Complex factors ranging from people's demographic background, personal experience, social circle, local contingency, and social and political institutions all work interactively to shape their opinions and perceptions of science (Miller 1999; Hansen 2009; Priest 2006).[2] At the same time, the book maintains that media representations are one important factor for science communication scholars to consider because media offer "a continuously changing cultural reservoir of images, meanings and definitions, on which different publics will draw for the purposes of articulating, making sense of and understanding science and the politics of science-based controversies and issues" (Hansen 2009, p. 118). In addition, "media accounts express relevant values and beliefs, help confer legitimacy to or discredit particular groups by treating them as part of the mainstream or as marginal, and therefore indirectly affect which perspectives do or do not ultimately come to dominate collective discourse and decision-making" (Priest 2006, p. 58). Put simply, though popular media do not determine public opinions, they attract reader attention, influence viewer perception, and "set the agenda" for social conversation (Miller 1999).

Visual Analysis Frameworks

To analyze popular science genetics images from the dynamic and multiple dimensions described above, the book draws upon two analytical frameworks: information design and social semiotics. The two frameworks are introduced below with additional details applied throughout the book.

Information Design

Information design is concerned with the graphic representation of complex and multimodal data, concepts, processes, and relationships in ways that encourage easy information retrieval and decision making. The term is closely related to, and often used interchangeably with, information visualization or simply visualization studies.[3] The framework has been used to examine diverse visual artifacts including scientific illustrations, technical diagrams, visual instructions, infographics, Cartesian graphs, and maps (see Tufte 1997; Tufte 2001; Kosslyn 2006; Neurath and Kinross 2009).

From a historical perspective, information design rose from the ancient human tradition of using images to record and convey information. In that regard, it predates verbal communication and manifests in such forms as European cave drawings, Egyptian tomb paintings, and Aztec codices. But more often, information design is traced to modern examples such as John Snow's 1854 spot map, which identifies a public water pump surrounded by cholera death as the source of contamination during London's cholera epidemic. Other well-known, earlier examples of information design include Florence Nightingale's 1858 rose diagram, which illustrates the causes and seasonal changes of patient mortality in the British military hospital during the Crimean War, and Charles Joseph Minard's 1861 diagram, which documents the movements and sizes of Napoleon's army as it invaded and retreated from Russia.

In its contemporary development, information design draws upon various theories in psychology and cognitive science. Notable among them is the Gestalt theory.[4] Established in the early twentieth century by a group of German psychologists, the theory explores how the human visual system perceives patterns out of visual stimuli. For example, the Gestalt principle of proximity states that, other things being equal, visual elements that are presented close together are perceived by viewers as conveying related information, regardless of their actual content; by contrast, elements that are far apart are perceived as conveying unrelated information, regardless of their actual content. Based on this understanding, information designers must consciously arrange data elements to correspond with their inherent logical relationships.

Also significant in information design is a general consensus on how the human sensory and cognitive system works to process visual input (Ware 2012; Kosslyn 1989). This process starts with viewers using their iconic memory to identify visual stimuli (e.g., lines, curves, or patches of color) sensed on the retina. From there, visual stimuli that are deemed interesting or relevant are processed by the working memory; this is where visual query and pattern search allow a viewer to recognize the stimuli as, say, a human face. The working memory also draws upon long-term memory, which includes prior knowledge and contextual information, so viewers can arrive at a pragmatic interpretation of the stimuli as, say, a friendly face. The second stage of this process and the role of the working memory are of particular interest to information design scholars, who maintain that humans' working memory has a rather low capacity and can "hold" only a few items at a given time (see Miller 1956; Kosslyn 1989; Ware 2012). When more varieties are introduced into a visual display, viewers are likely to make mistakes in visual processing. This understanding, in essence, drives many information design principles that emphasize simplicity, consistency, and contrast. For example, information designers recommend using a limited number of syntactic varieties in a visual display so as to reduce the demand on viewers' working memory. Or, they recommend a strong contrast between visual foreground and background so viewers' working memory can easily analyze salient visual patterns.

With its focus on perceptual and cognitive processing, information design provides a suitable framework for studying popular science genetics images: namely, whether and how these images are designed to convey scientific information to an audience who does not have extensive long-term memory and prior knowledge about the subject matter. But the relevance of the framework to this study goes further. Another, less-known aspect of information design concerns with "emotional interest," a concept made popular by Walter Kintsch and describes interest that occurs when "information evokes a strong affective response in the reader such as elation, disgust, or anger" (Schraw et al. 2001, p. 213). This concept enriches an otherwise purely cognitive process where readers derive satisfaction only from understanding the information presented.

Given this book's intention to examine popular science genetics images as both scientific and social–cultural artifacts, emotion and affective response is an important aspect to consider. As Priest (2009) put it, for everyday readers who do not share "the worldview of scientists in which knowledge (empirically validated) is an inherent good, much of science communication will capture audiences only if it also meets other needs—if it entertains as well as informs" (p. 230). Images that use rich syntactic details to arouse emotion (such as fear or wonder) are more likely to attract busy readers. Indeed, even some scientists consider affect (e.g., curiosity and enjoyment) to be the foundation of science (Li and Tsai 2013). On the other hand, how emotion and affect actually impact information processing is debatable. Conflicting findings exist on whether interesting but extraneous details interfere with or facilitate readers' comprehension (see Goetz and Sadoski 1995; Harp and Mayer 1997; Harp and Mayer 1998; Schraw 1998). More specifically with visual representation, experts disagree on whether minimalist design or artistic elaboration may be more appropriate for informational delivery (see Dragga and Voss 2001; Christiansen 2013).

The present study does not "side" with one or the other argument. It holds, as Schraw et al. (2001), that "interest" is not an isolated construct but complicated by various factors such as the nature of the information being processed, readers' prior knowledge, and the kind of "interesting" detail added. To illustrate these complexities, the book presents and compares visual examples that use or avoid affective details. It then discusses the different and multiple effects and motivations of these visual choices within the images' historical, scientific, and social–cultural contexts.

Social Semiotics

The information design framework as described above focuses on the apparent display of an image, but often, popular science genetics images belie implicit value stances and motives. This is where social semiotic analysis becomes relevant. Semiotics is widely acknowledged to have originated from the work of Swiss linguist Ferdinand de Saussure. According to Saussure, there are two parts to a sign: the signifier, which is a sound pattern (e.g., *pen* is pronounced/pɛn/), and the signified, which is the concept denoted by the sound pattern (e.g.,/pɛn/is a writing device). Although the original Saussurean concept treats both the signifier and signified as linguistic and non-material, contemporary semiotics generally acknowledges the materiality of signs (Chandler 2007). Thus, signifiers may include not only the spoken sound/pɛn/, but also the written word *pen,* or, of more relevance to this project, a visual representation of a pen. This development gave rise to the common definition of semiotics as the study of signs, where signs can be "anything which 'stands for' something else" and can "take the form of words, images, sounds, gestures and objects" (Chandler 2007, p. 2).

According to Saussure, linguistic signs are arbitrary: that is, there is no reason why we should call a pen/pɛn/; had we decided to call a pen/xɛn/, it would not change the signified of this sign. Based on this position, Saussure and the school of semiotics influenced by him (the structuralist semiotics) argue that a sign is not determined extralinguistically but based on intralinguistic rules; the focus of semiotics studies is thus the formal rules and conventions about linguistic syntax, lexicon, and phoneme (Chandler 2007). Again, later developments modified this original Saussurean model. To start, Jacques Lacan proposes that with a given sign, it is the signifier (e.g., the word *pen*), rather than the signified (e.g., the actual pen), that assumes primacy (Chandler 2007). This is because only by having a name for something can we then conceptualize and intelligently present or discuss it. In other words, reality is actively constructed rather than passively reflected by signs. Along this line of development, the post-structuralist semiotics or social semiotics refutes Saussure's concept of arbitrary signs and argues that a sign is not random or immutable but constructed and dynamic. As Kress and van Leeuwen (2006) put it, there is nothing natural about signs; the process of sign-making is driven by a sign-maker's interest. That interest "is a complex one, arising out of the cultural, social and psychological history of the sign-maker, and focused by the specific context in which the sign-maker produces the sign" (p. 7). Based on their interest and unique context, sign-makers choose to highlight certain aspects of an object, and those aspects, once fixed in a sign, can be accepted as an adequate, true representation of the object (Kress and van Leeuwen 2006).

Conceived this way, social semiotics helps to reveal the unarticulated value stance in everyday signs, including the seemingly commonsensical and mundane ones. Kress and van Leeuwen (2006), for example, derived insights from the contemporary use of rectangular shapes and argued that these shapes, with their parallel lines and controlled angles, give rise to our collective perception and performance of a rational, disciplined, and impersonal modern society:

> In contemporary Western society, squares and rectangles are the elements of the mechanical, technological order, of the world of human construction. They dominate the shape of our cities, our buildings, and our roads. They dominate the shape of many of the objects we use in daily life, including our pictures, which nowadays rarely have a round or oval frame, though other periods were happy to use these to frame more

intimate portraits in particular. Unlike circles, which are self-contained, complete in themselves, rectangular shapes can be stacked, aligned with each other in geometrical patterns: They form the modules, the building blocks with which we construct our world, and they are therefore the dominant choice of builders and engineers, and of those who think like builders and engineers. (p. 54)

Similar analyses, but of complex signs, are performed by Roland Barthes, who is best known for denaturalizing the petit-bourgeoisie's normalized signs (what he called myths) and revealing their political and ideological foundations. In Barthes' work, *Elle* magazine's beautiful photographs of dishes—"golden partridges studded with cherries, a faintly pink chicken chaudfroid, a mould of crayfish surrounded by their red shells, a frothy charlotte prettified with glacé fruit designs, multicoloured trifle, etc."—represent the petit-bourgeois preoccupation with ornamentation and their attempt to hide the real food and the economic cost of food (Barthes 1991, p. 78). These photographs "never show the dishes except from a high angle, as objects at once near and inaccessible, whose consumption can perfectly well be accomplished simply by looking. It is, in the fullest meaning of the word, a cuisine of advertisement, totally magical, especially when one remembers that this magazine is widely read in small-income groups" (Barthes 1991, p. 79). When *Elle*'s working class readers accept ornamental cookery as the norm and are more concerned with "sticking cherries into a partridge" (Barthes, p. 79) than having the partridge, they turn the petit-bourgeois myth into the version of reality that "matters."

Social semiotics as reviewed above serves the book's goal to study popular science genetics images as social–cultural artifacts, to reveal implicit and elusive motives (including political, economic, and disciplinary ones) behind media representations of genetics, and to demonstrate how media, in turn, participate in (re)constructing those motives and social realities. Although few studies have explicitly pursued such studies of popular science communication, those that do demonstrate the value of this analytical approach (see Mellor 2009; Anyfandi et al. 2014). Mellor (2009), in particular, examined media's portrayal of genetic modification and particle physics and demonstrated how decisions such as picture composition and camera angles can signify the concept of science.

It should be noted that semiotic analyses, whether in studying popular science communication or other contexts, are "inevitably personal and subjective" in order to "question what often goes unquestioned in the daily round of media production and consumption" (Mellor 2009, p. 218). This personal nature, however, does not mean that a researcher's interpretations are automatically unreliable, which is a positivist critique too often levied on qualitative research. In the case of semiotic analysis, a researcher "lays before readers the textual evidence to support the interpretation offered and allows them to judge for themselves how persuasive they find it" (Mellor 2009, p. 218). The ultimate goal of the analysis is not to make any claim to truth but to deconstruct otherwise black-boxed signs[5] and elicit critical conversations and additional research.

Combined Analysis

As reviewed above, information design and social semiotics have distinct assumptions: one prioritizes psychological and neurological processes, and the other emphasizes constructed interpretations. These two frameworks are thus often separately employed by visual communication scholars, leading to studies that focus either on best strategies to designing information (see Tufte 1997, 2001; Kosslyn 2006; Ware 2012) or socially grounded visual critique (see Brasseur 2003; Anker and Nelkin 2004; Kress and Leeuwen 2006).

This book purposefully combines the two frameworks to produce a more fluid and varied analysis, to exemplify the multi-dimensional approach to popular science communication promised earlier. With this combined analysis, the book discusses, on one hand, popular science genetics images' apparent appearance and affordance in information delivery, and on the other, their unapparent messages and implied emotions and value stances. Indeed, as we entertain such an analysis, the distinction between the two frameworks starts to become fuzzy: Information design includes affective consideration, which connects with semiotics' concern over personal and collective feelings, motives, and judgment; moreover, semiotic analysis is grounded in such syntactic details as lines, angles, and colors, which are the precise interest of information design.

Research Methods

Data Collection

Popular science magazines are the primary data source for this book. Compared with other popular media such as newspapers, general magazines, and science fiction titles, popular science magazines have the scope to provide more, and more in-depth, reports on genetics; they also have the tradition and space to use more, larger, higher-resolution, and full-color visual representations. Equally importantly, magazine archives retain century-old images, whereas newspaper archives, given their enormous daily publications, often have to forgo images. Popular science magazines thus allow the kind of longitudinal study pursued in this book: namely, how do popular science genetics images change as research focus, technology, and knowledge base change; as visualization software, hardware, and platform develop; and as social context, public interest, and political environment evolve.

Specifically, four magazines were used to source data: *American Scientist* (published since 1913), *Popular Science* (published since 1872), *Science News* (published since 1922), and *Scientific American* (published since 1845). These publications represent the best known popular science magazines in the USA and boast viewer bases in the millions (see, e.g., *Scientific American media kit* 2016; Society for Science and the Public 2016). As long-running publications, they also facilitate the book's longitudinal analysis. To locate genetics images from these sources, I first read a small sample of their articles that address topics in genetics and identified a series of keywords: *DNA, gene, genetic, genetics, genome, genomic, genomics, heredity,* and *inheritance.* These keywords were then searched within the respective magazine archives. All resultant articles were reviewed. Articles that do not actually focus on genetics, despite containing the keywords, were eliminated; also eliminated were articles that do not use pertinent visual representations. This resulted in 890 articles, which contain around 5100 images. The selections span over 100 years, with the earliest published in 1905 and the latest ones in 2016, the time this book finished writing.

In addition to popular science magazine images, the book references other high-profile, public-facing genetics images to further situate and support its discussion. These include images publicized by the National Human Genome Research Institute, the U.S. Food and Drug Administration, and the National Center for Biotechnology Information. Images were found in the "outreach," "consumers," "education," and media gallery sections of these organizations' Web sites. Also referenced are images published in national newspapers (such as *The New York Times* and *Washington Post*), which were located by searching the newspapers' image archives. These additional images are generally published within the last decade and help substantiate the book's discussion of contemporary popular science communication, in particular, the use of digital, interactive formats.

Data Analysis

Images were individually examined and notes were kept on their year of publication; visual genre (e.g., illustration, photograph, etc.); content; and features that pertain to the two frameworks reviewed earlier (information design and social semiotics). For example, notes for one image may state that it was published in 2010, is a procedural illustration on the process of protein translation, is eye-catching with abundant color and three-dimensional objects, and is relatively easy to access given its use of labels and lack of jargon.

The analysis of the images and notes followed the constant comparative method (Glaser 1965; Merriam 2007). That is, as substantial notes were made, they were compared so their commonalities could give rise to tentative themes. The above notes, for example, contributed to the genre theme of "illustrations" and more specifically, the theme of "syntactically complex illustrations." Throughout the analysis, tentative themes were modified, integrated, or separated based on a continuous comparison of visuals that belonged in the same and related themes. During this comparison, memos were also kept to record emerging ideas on particular aspects of the themes. For example, examining illustrations from different time periods led to the observation that contemporary ones are more likely to be syntactically complex than those published prior to the 1980s.

The study's data analysis and resultant findings are thus inductive and emergent and based on the researcher's reflective interpretation of the data (Glaser and Strauss 1967; Hardy et al. 2004; Merriam 2007). Rather than trying to record the frequency of certain visual elements or to derive a quantitative meaning, the analysis explores how certain discourse representations are made possible, plausible, or "taken for granted" and how they articulate and shape social perceptions (Laffey and Weldes 2004). This process serves the purpose of the book: to broach popular science genetics images as flexible and multidimensional artifacts of science and society.

Data Overview

Chapters in this book are organized based on visual genres (more about this below), but in the following, I also provide an overview of the visual data based on broad time periods for readers who are interested in historic overviews.

1900s to mid-1940s: The number of popular science genetics images published in these decades is relatively small. This is not surprising given that the discipline was young and had not generated widespread public interest. In her study of public perception of genetics, van Dijck (1998) omitted this entire period. There was, of course, substantial public discussion of eugenics during the time (see Condit 1999), but this study did not find notable visual representations of eugenics beyond those on the selective breeding of plants and animals. Indeed, images used in this period mostly include photographs and illustrations that portray the appearance of plants and animals.

Mid-1940s to 1950s: In 1944, DNA was established as the chemical basis of organisms' characteristics (see Avery et al. 1944). DNA's double-helix structure was subsequently revealed in 1953 by James Watson and Francis Crick. Reflecting this series of groundbreaking work, the amount of media reporting on genetics and the number of relevant images dramatically increased. Notably, micrographs and structural formulas that emphasize, respectively, the physical and molecular composition of gene products appeared. As DNA entered the media spotlight, images that revolve around the DNA code also appeared. During this period, limited colors started to be employed in visual representations.

1960s to mid-1980s: The number of popular science genetics images continued to increase during these decades, reflecting the growing and increasingly more complex research being pursued by scientists, for example, the formation of chromosomes, the effect of transposable genetic elements, or the biosynthesis of RNA. Notably, Cartesian graphs became common in the 1960s, corresponding to the increasingly quantitative nature of genetic research. The use of color also increased but remained modest.

Mid-1980s to the present: Maintaining the momentum of earlier decades, mass media reporting and visual representations of genetics continue to be strong. Compared with earlier works, there is less focus on genetics as "pure" research and more on its social, environmental, and especially medical implication and application: for example, the controversy surrounding genetically modified food or the promise of genetic medicine. Images from these decades regularly use full colors, are more elaborate, and are higher in visual resolution. In addition, the development of visualization technologies gave rise to a variety of digital, interactive formats including games, applets, and customizable database tools.

Chapter Organization

The remainder of the book, except for the conclusion chapter, is organized based on one coding theme: visual genre. Fairly early in the data analysis process, it became clear that the collected images can be meaningfully categorized into genres to reflect their shared characteristics and development. By genre, I mean, as genre scholars generally do, typified (visual) discourses that respond to recurrent rhetorical situations; these discourses are subject to change with changed circumstances and reflect as well as participate in constructing social realities (see Miller 1984; Berkenkotter and Huckin 1995; Bazerman 2010). For example, micrographs emerged as a visual genre when they responded to the typified need of revealing minute organelles and phenomena. Micrographs' function and significance change as microscopic technologies develop and scientific paradigms shift, and their use reflects as well as enables cellular and molecular research.

It is important to note that genre does not equate to shared syntactic or stylistic appearance. As Miller (1984) argued, reducing genres to surface-level forms creates a closed system that sacrifices the diversity and dynamism of the genre concept. A case in point is the visual representation of the genetic code. Over the past 50 years, a range of images (from tables to sequence reads) have been used in popular media to extrapolate this grand metaphor. Although these images are apparently dissimilar, they have shared social functions and respond to the same rhetorical exigency: to try to give shape to an abstract but central premise in genetics. Their surface-level difference is merely a result of changed disciplinary, technological, and social contexts—just as a nineteenth-century fuzzy, black-and-white optical micrograph looks different from today's brightly colored, extremely magnified scanning electron micrographs. With the code images, preoccupation with surface forms would have missed insights in their evolving construction and interpretation.

With the above conception, six visual genres emerged from the data and gave rise to the following chapters: Chap. 2 focuses on the use of photographs, Chap. 3 discusses micrographs, Chap. 4 examines illustrations, Chap. 5 considers visuals that revolve around the genetic code, Chap. 6 examines quantitative graphs, and Chap. 7 discusses images that reveal the molecular nature of genetics. Within each chapter, the discussion is organized around significant genre features and functions as well as historical trajectories.

Notes

1. In the traditional view, public readers are often conceived as a homogeneous singular entity, "the public."
2. Examining such personal factors and their relationship vis-à-vis science visual communication would be a topic for another book.
3. Historically, visualization studies may be categorized into information visualization and scientific visualization, with the former supposedly representing abstract, non-physical data and the latter representing physical, spatial data (Rhyne 2003). But this division fails to account for data (such as genomics data) that do not fit neatly into either category; more generally, this division is increasingly recognized as artificial, and it falsely suggests that scientific visualization is not informative or information visualization is not scientific (Rhyne 2003; Johnson 2004).
4. *Gestalt* is German for *pattern*.
5. As Latour (1998) would call them.

References

Aldahmash, A. H., & Abraham, M. R. (2009). Kinetic versus static visuals for facilitating college students' understanding of organic reaction mechanisms in chemistry. *Journal of Chemical Education, 86*(12), 1442–1446.

Amann, K. & Knorr-Cetina, K. (1990). The fixation of (visual) evidence. In M. Lynch & S. Woolgar (Eds.), *Representation in scientific practice* (pp. 85–121). Cambridge, MA: The MIT Press.

Anker, S., & Nelkin, D. (2004). *The molecular gaze: Art in the genetic age.* Cold Spring Harbor, NY: Cold Spring Harbor Laboratory Press.

Anyfandi, G., Koulaidis, V., & Dimopoulos, K. (2014). A socio-semiotic framework for the analysis of exhibits in a science museum. *Semiotica, 2014*(200), 229–254. doi:10.1515/sem-2014-0001.

Avery, O. T., Macleod, C. M., & McCarty, M. (1944). Studies on the chemical nature of the substance inducing transformation of pneumococcal types: Induction of transformation by a desoxyribonucleic acid fraction isolated from pneumococcus type III. *The Journal of Experimental Medicine, 79*(2), 137–158.

Barthes, R. (1991). *Mythologies.* New York: Noonday Press.

Bazerman, C. (2010). Rhetorical genre studies. In A. S. Bawarshi & M. J. Reiff (Eds.), *Genre: An introduction to history, theory, research, and pedagogy* (pp. 78–104). West Lafayette, IN: Parlor Press and The WAC Clearinghouse.

BBVA Foundation. (2012). BBVA foundation international study on scientific culture: Understanding of science. Retrieved July 21, 2017, from http://w3.grupobbva.com/TLFU/dat/Understandingsciencenotalarga.pdf.

Berkenkotter, C., & Huckin, T. (1995). *Genre knowledge in disciplinary communication: Cognition/culture/power*. Mahwah, NJ: Lawrence Erlbaum.

Blum, A. S. (1993). *Picturing nature: American nineteenth-century zoological illustration*. Princeton, NJ: Princeton University Press.

Brasseur, L. E. (2003). *Visualizing technical information: A cultural critique*. Amityville, NY: Baywood.

Cambrosio, A., Jacobi, D., & Keating, P. (1993). Ehrlich's "beautiful pictures" and the controversial beginnings of immunological imagery. *Isis, 84*(4), 662–699.

Chandler, D. (2007). *Semiotics: The basics*. London: Routledge.

Chittleborough, G., & Treagust, D. (2008). Correct interpretation of chemical diagrams requires transforming from one level of representation to another. *Research in Science Education, 38*(4), 463–482.

Christiansen, J. (2013). A defense of artistic license in illustrating scientific concepts for a non-specialist audience. In *Communicating Complexity 2013 Conference Proceedings* (pp. 49–60). Rome: Edizioni Nuova Cultura-Roma.

Condit, C. M. (1999). *The meanings of the gene: Public debates about human heredity*. Madison: University of Wisconsin Press.

Cook, M., Wiebe, E. N., & Carter, G. (2008). The influence of prior knowledge on viewing and interpreting graphics with macroscopic and molecular representations. *Science Education, 92*(5), 848–867.

Davidson, J. P. (2008). *A history of paleontology illustration*. Bloomington: Indiana University Press.

Dinolfo, J., Heifferon, B., & Temesvari, L. A. (2007). Seeing cells: Teaching the visual/verbal rhetoric of biology. *Journal of Technical Writing and Communication, 37*(4), 395–417.

Dragga, Sam, & Voss, Dan. (2001). Cruel pies: The inhumanity of technical illustrations. *Technical Communication, 48*(3), 265–274.

Durodié, B. (2003). Limitations of public dialogue in science and the rise of new 'experts'. *Critical Review of International Social and Political Philosophy, 6*(4), 82–92.

Falk, J. H., Storksdieck, M., & Dierking, L. D. (2007). Investigating public science interest and understanding: Evidence for the importance of free-choice learning. *Public Understanding of Science, 16*(4), 455–469. doi:10.1177/0963662506064240.

Ford, B. J. (1993). *Images of science: A history of scientific illustration*. New York: Oxford University Press.

Galison, P. (1998). Judgment against objectivity. In C. A. Jones, P. Galison, & A. E. Slaton (Eds.), *Picturing science, producing art* (pp. 327–359). New York: Routledge.

Glaser, B. G. (1965). The constant comparative method of qualitative analysis. *Social Problems, 12*(4), 436–445.

Glaser, B. G., & Strauss, A. (1967). *The discovery of grounded theory: Strategies for qualitative research*. Chicago: Aldine Publishing Company.

Grand, A. (2009). Engaging through dialogue: International experiences of Café Scientifique. In R. Holliman, J. Thomas, S. Smidt, E. Scanlon, & E. Whitelegg (Eds.), *Practicing science communication in the information age* (pp. 209–226). Oxford: Oxford University Press.

Goetz, E. T., & Sadoski, M. (1995). Commentary: The perils of seduction: Distracting details or incomprehensible abstractions? *Reading Research Quarterly, 30*(3), 500–511.

Gould, S. J. (1993, October). Dinosaur deconstruction. *Discover, 14*, 108–113.

Hansen, A. (2009). Science, communication and media. In R. Holliman, E. Whitelegg, E. Scanlon, S. Smidt, & J. Thomas (Eds.), *Investigating science communication in the information age* (pp. 105–127). Oxford: Oxford University Press.

Hardy, C., Harley, B., & Phillips, N. (2004). Discourse analysis and content analysis: Two solitudes. *Qualitative Methods, 2*(1), 19–22.

Harp, S. F., & Mayer, R. E. (1997). The role of interest in learning from scientific text and illustrations: On the distinction between emotional interest and cognitive interest. *Journal of Educational Psychology, 89*(1), 92–102.

Harp, S. F., & Mayer, R. E. (1998). How seductive details do their damage: A theory of cognitive interest in science learning. *Journal of Educational Psychology, 90*(3), 414–434.

Hildebrand, R. (2004). Alternative images: Anatomical illustration and the conflict between art and science. *Interdisciplinary Science Reviews, 29*(3), 295–311.

Holliman, R., & Jensen, E. (2009). (In)authentic sciences and (im)partial publics: (Re)constructing the science outreach and public engagement agenda.

In R. Holliman, E. Whitelegg, E. Scanlon, S. Smidt, & J. Thomas (Eds.), *Investigating science communication in the information age* (pp. 35–52). Oxford: Oxford University Press.

Holliman, R., Whitelegg, E., Scanlon, E., Smidt, S., & Thomas, J. (2009a). *Investigating science communication in the information age: Implications for public engagement and popular media.* Oxford: Oxford University Press.

Holliman, R., Thomas, J., Smidt, S., Scanlon, E., & Whitelegg, E. (2009b). *Practising science communication in the information age: Implications for public engagement and popular media.* Oxford: Oxford University Press.

Irwin, A., & Michael, M. (2003). *Science, social theory & public knowledge.* Philadelphia: Open University Press.

Irwin, A., & Wynne, B. (Eds.). (2004). *Misunderstanding science? The public reconstruction of science and technology.* Cambridge: Cambridge University Press.

Johnson, C. (2004). Top scientific visualization research problems. *Computer Graphics and Applications, IEEE, 24*(4), 13–17. doi:10.1109/MCG.2004.20.

Kemp, M. (1970). A drawing for the *Fabrica*; and some thoughts upon the Vesalius muscle-men. *Medical History, 14*(3), 277–288.

Kemp, M. (2000). *Visualizations: The nature book of art and science.* Oakland: University of California Press.

Knorr-Cetina, K., & Amann, K. (1990). Image dissection in natural scientific inquiry. *Science, Technology and Human Values, 15*(3), 259–283.

Kosslyn, S. M. (1989). Understanding charts and graphs. *Applied Cognitive Psychology, 3*(3), 185–225.

Kosslyn, S. M. (2006). *Graph design for the eye and mind.* New York: Oxford University Press.

Kress, G. R., & Van Leeuwen, T. (2006). *Reading images: The grammar of visual design* (2nd ed.). London: Routledge.

Krzywinski, M., Schein, J., Birol, I., Connors, J., Gascoyne, R., Horsman, D., … Marra, M. A. (2009). Circos: An information aesthetic for comparative genomics. *Genome Research, 19*(9), 1639–1645.

Laffey, M., & Weldes, J. (2004). Methodological reflections on discourse analysis. *Qualitative Methods, 2*(1), 28–30.

Latour, B. (1986). Visualization and cognition: Thinking with eyes and hands. In H. Kuklick & E. Long (Eds.), *Knowledge and society: Studies in the sociology of culture past and present* (Vol. 6, pp. 1–40). Greenwich, CT: Jai Press.

Latour, B. (1998). How to be iconophilic in art, science, and religion. In C. A. Jones, P. Galison, & A. E. Slaton (Eds.), *Picturing science, producing art* (pp. 418–440). New York: Routledge.

Latour, B., & Woolgar, S. (1979). *Laboratory life: The construction of scientific facts*. Princeton, NJ: Princeton University Press.

Li, M., & Tsai, C. (2013). Game-based learning in science education: A review of relevant research. *Journal of Science Education and Technology, 22*(6), 877–898. doi:10.1007/s10956-013-9436-x.

Lynch, M. (1985). *Art and artifact in laboratory science: A study of shop work and shop talk in a research laboratory*. London: Routledge & Kegan Paul.

Lynch, M. (1990). The externalized retina: Selection and mathematization in the visual documentation of objects in the life sciences. In M. Lynch & S. Woolgar (Eds.), *Representation in scientific practice* (pp. 153–186). Cambridge, MA: MIT Press.

Lynch, M., & Woolgar, S. (1990). *Representation in scientific practice*. Cambridge, MA.: MIT Press.

MacDonald, S. (2004). Authorising science: Public understanding of science in museums. In A. Irwin & B. Wynne (Eds.), *Misunderstanding science? The public reconstruction of science and technology* (pp. 152–171). Cambridge: Cambridge University Press.

Merriam, S. B. (2007). *Qualitative research and case study applications in education*. San Francisco, CA: Jossey-Bass.

McCormick, B. H., DeFanti, T. A., & Brown, M. D. (1987). Visualization in scientific computing. *Computer Graphics, 21*(6), i–E8.

Meisner, R., & Osborne, J. (2009). Engaging with interactive science exhibits: A study of children's activity and the value of experience for communicating science. In R. Holliman, E. Whitelegg, E. Scanlon, S. Smidt, & J. Thomas (Eds.), *Investigating science communication in the information age* (pp. 86–102). Oxford: Oxford University Press.

Mellor, F. (2009). Image-music-text of popular science. In R. Holliman, E. Whitelegg, E. Scanlon, S. Smidt, & J. Thomas (Eds.), *Investigating science communication in the information age* (pp. 205–220). Oxford: Oxford University Press.

Michael, M. (2002). Comprehension, apprehension, prehension: Heterogeneity and the public understanding of science. *Science, Technology and Human Values, 27*(3), 357–378.

Miller, G. A. (1956). The magical number seven, plus or minus two: Some limits on our capacity for processing information. *Psychological Review, 63*(2), 81–97.

Miller, C. R. (1984). Genre as social action. *Quarterly Journal of Speech, 70*(2), 151–167. doi:10.1080/00335638409383686.

Miller, D. (1999). Mediating science: Promotional strategies, media coverage, public belief and decision making. In E. Scanlon, E. Whitelegg, & S. Yates (Eds.), *Communicating science: Contexts and channels* (pp. 206–226). London: Routledge.

Miller, J. D. (2004). Public understanding of, and attitudes toward, scientific research: What we know and what we need to know. *Public Understanding of Science, 13*(3), 273–294.

Morgan, D. H., Kristensen, D. M., Mittelman, D., & Lichtarge, O. (2006). ET viewer: An application for predicting and visualizing functional sites in protein structures. *Bioinformatics, 22*(16), 2049–2050. doi:10.1093/bioinformatics/btl285.

National Science Board. (2014). *Science and engineering indicators 2014.* (No. NSB 14-01). Arlington, VA: National Science Foundation.

Neurath, M., & Kinross, R. (2009). *The transformer: Principles of making isotype charts.* London: Hyphen Press.

Nisbet, M. C., & Scheufele, D. A. (2009). What's next for science communication? Promising directions and lingering distractions. *American Journal of Botany, 96*(10), 1767–1778. doi:10.3732/ajb.0900041.

Northcut, K. (2006). Images as facilitators of public participation in science. *Journal of Visual Literacy, 26*(1), 1–14.

Patrick, M. D., Carter, G., & Wiebe, E. N. (2005). Visual representations of DNA replication: Middle grades students' perceptions and interpretations. *Journal of Science Education and Technology, 14*(3), 353–365.

Pauwels, L. (2006). *Visual cultures of science: Rethinking representational practices in knowledge building and science communication.* Lebanon, NH: Dartmouth College.

Pintó, R., & Ametller, J. (2002). Students' difficulties in reading images. Comparing results from four national research groups. *International Journal of Science Education, 24*(3), 333–341.

Pozzer, L., & Roth, W. (2003). Prevalence, function, and structure of photographs in high school biology textbooks. *Journal of Research in Science Teaching, 40*(10), 1089–1114.

Pozzer-Ardenghi, L., & Roth, W. (2004). Making sense of photographs. *Science Education, 89*(2), 219–241.

Priest, S. H. (2006). The public opinion climate for gene technologies in Canada and the United States: Competing voices, contrasting frames. *Public Understanding of Science, 15*(1), 55–71. doi:10.1177/0963662506052889.

Priest, S. H. (2009). Reinterpreting the audiences for media messages about science. In R. Holliman, E. Whitelegg, E. Scanlon, S. Smidt, & J. Thomas (Eds.), *Investigating science communication in the information age* (pp. 223–236). Oxford: Oxford University Press.

Rhyne, T. (2003). Does the difference between information and scientific visualization really matter? *Computer Graphics and Applications, IEEE, 23*(3), 6–8. doi:10.1109/MCG.2003.1198256.

Roth, W., Pozzer-Ardenghi, L., & Han, J. Y. (2005). *Critical graphicacy: Understanding visual representation practices in school science.* Dordrecht: Springer.

Rundgren, C., & Tibell, L. A. E. (2009). Critical features of visualizations of transport through the cell membrane: An empirical study of upper secondary and tertiary students' meaning-making of a still image and an animation. *International Journal of Science and Mathematics Education, 8*(2), 223–246.

Schraw, G. (1998). Processing and recall differences among seductive details. *Journal of Educational Psychology, 90*(1), 3–12.

Schraw, G., Flowerday, T., & Lehman, S. (2001). Increasing situational interest in the classroom. *Educational Psychology Review, 13*(3), 211–224. doi:10.10 23/A:1016619705184.

Scientific American Media Kit. (2016). Retrieved February 24, 2016, from https://www.scientificamerican.com/mediakit/.

Society for Science & the Public. (2016). *Science News.* Retrieved February 24, 2016, from https://www.societyforscience.org/science-news.

Stilgoe, J., & Wilsdon, J. (2009). The new politics of public engagement with science. In R. Holliman, E. Whitelegg, E. Scanlon, S. Smidt, & J. Thomas (Eds.), *Investigating science communication in the information age* (pp. 18–34). Oxford: Oxford University Press.

Stylianidou, F., Ormerod, F., & Ogborn, J. (2002). Analysis of science textbook pictures about energy and pupils' readings of them. *International Journal of Science Education, 24*(3), 257–283.

Thomas, J. (2009). Controversy and consensus. In R. Holliman, J. Thomas, S. Smidt, E. Scanlon, & E. Whitelegg (Eds.), *Practicing science communication in the information age* (pp. 131–148). Oxford: Oxford University Press.

Trumbo, J. (2000). Seeing science research opportunities in the visual communication of science. *Science Communication, 21*(4), 379–391.

Tufte, E. R. (1997). *Visual explanation: Images and quantities, evidence and narrative.* Cheshire, CT: Graphics Press.

Tufte, E. R. (2001). *The visual display of quantitative information*. Cheshire, CT: Graphics Press.

van Dijck, J. (1998). *Imagenation: Popular images of genetics*. New York: New York University Press.

van Dijck, J. (2003). After the "Two cultures": Toward a "(multi)cultural" practice of science communication. *Science Communication, 25*, 177–190.

Walter, T., David, W. S., Baldock, R., Mark, E. B., Anne, E. C., Duce, S., ... Hériché, J. (2010). Visualization of image data from cells to organisms. *Nature Methods, 7*(3), S26–S55.

Ware, C. (2012). *Information visualization: Perception for design* (3rd ed.). Burlington, MA: Morgan Kaufmann.

Wickman, C. (2013). Observing inscriptions at work: Visualization and text production in experimental physics research. *Technical Communication Quarterly, 22*(2), 150–171.

Wynne, B. (2004). Misunderstood misunderstandings: Social identities and public uptake of science. In A. Irwin & B. Wynne (Eds.), *Misunderstanding science? The public reconstruction of science and technology* (pp. 19–46). Cambridge: Cambridge University Press.

2

The Photographic View: Observational Record and Symbolic Excess

Photographs are a common visual genre used by popular science magazines and other mass media to communicate genetics to public readers. Their use spans the entire history of modern genetics and continues to evolve as the discipline and its social context change. As a genre familiar to public readers and capable of displaying concrete visual details, photographs, as this chapter argues, can make for relatable and accessible visual evidence. At the same time, photographs' rich (some would say excess) visual details can also mean a lack of semantic determinacy; that is, their meanings are subject to the potentially different interpretations of a viewer (Pozzer-Ardenghi and Roth 2004). Examining photographs' evolving and multiple roles in the popular communication of genetics, this chapter sheds light on how this visual genre partakes in shaping public perceptions of genetics. In particular, the chapter emphasizes two interrelated functions of photographs: as primarily informative, cognition-based evidence and as primarily symbolic, affect-based artifacts.

© The Author(s) 2017
H. Yu, *Communicating Genetics*,
DOI 10.1057/978-1-137-58779-4_2

Informative Photographs: Evidence from Classical Genetics

It is, by now, a well-known story. Gregor Johann Mendel (1822–1884), posthumously revered the father of modern genetics, was a little-known Austrian monk in his lifetime. While trying to create hybrid pea plants in his monastery garden, he established what is known today as the Mendel's Law of Inheritance. Put somewhat simplistically, Mendel's Law states that alleles (multiple variants of one same gene) are separately and randomly passed from parents to offspring; the combination of the alleles and their dominant or recessive state determines the appearance of the offspring. Mendel published his findings in 1866 in *Versuche über Pflanzen-Hybriden* (*Experiments on Plant Hybridization*); it was, however, not until the early twentieth century when his work became widely known and confirmed by other researchers, which ushered in the roughly 40-year span of classical genetics.

During this era, photographs were used to demonstrate the apparent results of inheritance and mutations in plants, trees, and animals (including humans). In particular, they were used to demonstrate how Mendel's Law can be employed for selective crossbreeding, as shown in Fig. 2.1, in the case of corn. On the left panel of Fig. 2.1 are two different corn strains, separated in two rows. These corns show obvious

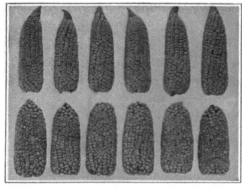

Fig. 2.1 Hybrid corn grows to superior size (Jones 1919, p. 230)

inter-strain consistency and cross-strain difference in their shape and structure. The right panel displays three corns. The smaller ones on the sides are the same two strains shown on the left panel; the larger one in the middle is their crossbred offspring. As the photograph makes clear, the crossbred strain has a larger size unmatched by its parents.

Similar success was gained in the breeding of tobacco. In Fig. 2.2 left, we see the Havana wrapper tobacco grown in the Connecticut River Valley. It features short, wind-resistant stalk and large leaves, but the leaves are moderate in number (averaging 19–21 per plant). Shown in Fig. 2.2 middle is the Cuban tobacco that has tall, non-wind-resistant stalk and medium-sized leaves, but the leaves are superior in number (averaging 26 per plant). By crossing the two breeds and recombining crops with desired features, one can, in the second generation, breed the Halladay shown in Fig. 2.2 right. It combines Havana's strong stalk and large-sized leaves with Cuban's superior number of leaves, leading to significantly higher yields. Photographic evidence similar to that shown in Figs. 2.1 and 2.2 was used to report genetics-based experiments and practice in animal breeding (e.g., Castle 1905) and later forestry development (e.g., Hunter 1951).

Fig. 2.2   Halladay tobacco outperforms its parents in stalk strength, leave size, and leave number (East 1910, pp. 350–352)

These early twentieth-century photographs may seem unsophisticated compared to today's high-tech visual splendor, but their perceived ability to photographically record "reality," especially through comparison, makes them powerfully persuasive (Dobrin and Morey 2009, p. 293). The comparison of different specimens illustrates, simultaneously, the experiment process (crossbreeding two strains), its outcomes (obtaining a hybrid), and its significance (a superior hybrid). As such, not only do the photographs explain scientific experiments, they are arguments for genetics-based agricultural practice and its social benefits.

These arguments were especially significant in the early twentieth-century America—when rapid urbanization and industrialization were taxing the agriculture industry and changing the agriculture paradigm "from widespread subsistence farming to a system of farms providing food for the newly urbanized areas" (Fulton 1998). As rural population migrated to cities and immigrants arrived, significantly fewer farmers were operating on fewer acres to feed more people (Fulton 1998). When America joined World War I in 1917, the food needs of its European allies also added to the demand (Fulton 1998). In this historical context, higher production yields, shorter growth periods, and an overall control over agricultural practice became a social urgency.

Early genetics photographs not only responded to this urgency by presenting scientific solutions but also made the solution a matter of personal pride. Exuding from these photographs is a sense of human triumph over Mother Nature. As an article on tree breeding exclaimed, "MAN IS now making new kinds of trees. The old, established kinds don't suit him any more. They grow too slowly, or they have too many branches, or they succumb too easily to disease or drought. So man has started to create trees which will not have such failing" (Hunter 1951, p. 10). Accompanying the statement is the same proud photograph, wherein the newly created hybrid trees are compared with the "old kinds" and show faster growth. Certainly, the invocation of "science triumphing over nature" is not particularly unique to these photographs. Media reports of genetics portray the same sentiment through word choices (Hansen 2006), and indeed, it is an ideology going back to the Scientific Revolution in Western culture (Cohen 1994). Yet photographs, in their supposed ability to capture nature "as is" and to fix a graphic moment in history, can be an especially convincing testimony to the claim.

Moreover, being significant social artifacts did not preclude these photographs from serving as well adapted scientific evidence. The early twentieth-century natural scientists were preoccupied with machine-made mechanical images, images that act as transparent conduits to present nature as if it were speaking for itself (Galison 1998). Within this paradigm, the machine was trusted "as a neutral and transparent operator that would serve both as instrument of registration without intervention and as an ideal for the moral discipline of the scientists themselves" (Galison 1998, p. 332). Photographs were thus the scientific visual evidence par excellence, for their creation was, supposedly, a mechanical and automatic process accomplished by none other than the objective camera—provided that any retouching or post-processing was prohibited.

This, of course, was a naïve view of photographs. Photography, or other machine-based visualization, is far from being an automatic process void of human interaction. As Meyer (2007) wrote, a photographer has ample opportunities to influence the outcome of a photograph, even without obvious retouching: "The position where the photographer stands in relation to the scene, the instant at which the exposure is made, the choice of camera, lens, shutter speed and aperture, and the selection of which photographs among many to print and publish" (p. 102). Furthermore, from a science communication perspective, photographs that are automated and free of retouching are not necessarily preferable. Without these "interventions," the excessive visual details of a photograph can obscure significant clues from non-expert readers (Pozzer-Ardenghi and Roth 2004). It is thus around the last two-thirds of the twentieth century that scientists in various disciplines came to celebrate the interpreted images, which are created by trained experts who, having observed and critically assessed many visual examples, synthesized the observation to bring out their presumably universal patterns and salient points for less-experienced viewers (Galison 1998). In the popular communication of genetics, this shift contributed to the illustrative capture of electrophoresis and the rise of symbolic photographs discussed later in this chapter.

Of course, conventional photographs have not disappeared and continue to be used as informative evidence—though their function is undergoing both subtle and obvious changes. As modern genetic experiments replaced Mendelian crossbreeding, photographs are used to document the physical appearance of organisms that are subject to various testing. Figure 2.3, for instance, is used in a *Science News* article (Brownlee 2006, p. 393) to demonstrate the effect of epigenetics, an area of study that looks beyond genes as the sole determinant of life and examines how external factors (environment, diet, etc.) exert control by influencing which genes are expressed. As Fig. 2.3 shows, mice that carry the same gene (*agouti*), which affects their fur color, can exhibit a range of colors and weights depending on the food their pregnant mothers were fed.

As earlier photographs, Fig. 2.3 demonstrates, from a visual perspective, the experiment results: that is, the mice assume varying fur colors and sizes. But compared with Figs. 2.1 and 2.2, it is less capable of demonstrating the experiment process in question or its significance. The ways by which the mice's colors and sizes were manipulated cannot be discerned from the photograph alone. In addition, as it turns out, these physical differences are not the most meaningful findings

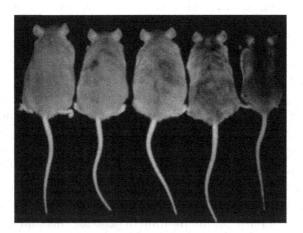

Fig. 2.3 Effects of epigenetics on mice fur color and size. Reproduced with permission from Dolinoy et al. (2006, p. 569) (color figure online)

either; what is of significance is that the brown and smaller mice have decreased risks for diseases such as diabetes and cancer, because their mothers were fed nutrients derived from soybean (Brownlee 2006).

By this, I am not suggesting that images like Fig. 2.3 are no longer valuable in popular communication. They are: They convey observable information and serve as evidence of contemporary research. What I am calling attention to is that as the scientific paradigm shifts, so does the function and relevance of particular visual genres. Classical genetics focused its inquiry on visible phenomena, which made photographs convenient and highly pertinent visual evidence. This research context changed in the 1950s: Experiments confirmed DNA's central role as genetic materials, and DNA's structure was subsequently determined. These breakthroughs brought about the DNA age of modern genetics and changed the discipline's focus from observable features to molecular-level activities. Because of this, photographs that depict objects and phenomena directly observable to the human eyes lost their prior advantage.[1]

What advantage they lost in conveying molecular information, however, the photographs gained in making genetics more "fun" and relatable for public audiences. Consider, for example, a photograph of a playful tiger and fish in a study that explores the molecular mechanisms behind the formation of animal skin/fur patterns (Saey 2010), or photographs of cuddly polar bears in a study that examines the bear's ancestral origin (Millius 2011), or a close-up of a beautiful monarch butterfly in an article that ponders the environmental effects of genetically modified crops (Brown 2001). These photographs do not try to function as scientific evidence the way Figs. 2.1, 2.2 and 2.3 do. Instead, they frame genetics in entertaining or at least familiar contexts for public readers. As such, they contribute to attracting reader attention, generating emotional interest, and developing a broader and arguably more socially, culturally, and environmentally grounded understanding of genetics.

Informative Photographs: DNA Fingerprinting

Although traditional photographs lost some of their advantages as direct evidence when genetics entered the DNA age, there is one important exception: the photographic capture of electrophoresis. Electrophoresis may not be a term familiar to public readers, but it is behind a technique often heard in mass media: DNA fingerprinting, which is commonly used in crime scene investigation, body identification, and paternal testing. More generally, electrophoresis is an analytical tool used to separate DNA, RNA, and protein molecules of different sizes.

Put simply, in electrophoresis, test samples such as a DNA chain is cut into fragments by enzymes. The fragments are loaded into some gel, which is then placed in an electric field. Because molecules are themselves electrically charged, they migrate in the gel. DNA, for example, is negatively charged so it migrates toward the positive end of the gel. Smaller fragments of DNA molecules move faster than larger ones, so given time, fragments of different sizes will travel different distances and form separate "bands." Agents and dyes can then be added to bind with the molecules and "stain" the bands, making them visible. The bands are captured via photographic techniques, including conventional photography and autoradiography.

Figure 2.4 provides one example, where electrophoresis demonstrates the unique reproductive practice of the nine-banded armadillo: A female typically gives birth to a litter of four genetically identical offspring, that is, a set of same-sex quadruplets. The left part of Fig. 2.4 shows electrophoresis results of four unrelated litters, separated by black vertical lines. Within each litter are four lanes, each representing one of the quadruplets. As can be seen, the four siblings within each litter have identical bands or the so-called DNA fingerprints; across litters, the bands differ. By contrast, the right part of the figure shows electrophoresis results of a great many random armadillos, who all display different DNA fingerprints.

With Fig. 2.4, readers get to "see" the abstract, molecular-level genetic research in more concrete terms. Even though the photograph itself is

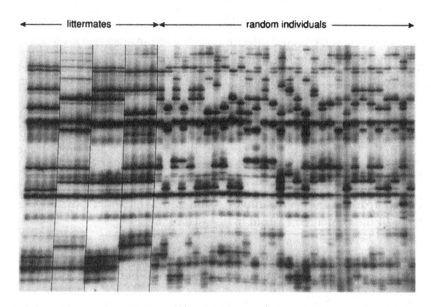

Fig. 2.4 Electrophoresis demonstrates nine-banded armadillos having quadruplet litters (Loughry et al. 1998, p. 278). Courtesy of William James Loughry and Paulo Prodöhl

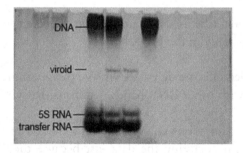

Fig. 2.5 Electrophoresis reveals that viroid is a type of RNA. Adapted by permission from Macmillan Publishers Ltd: Nature Reviews Microbiology (Diener 2003, p. 77), copyright 2003

incapable of explaining the full technique of genetic fingerprinting, it helps to provide a frame of reference as to what is being observed and examined.

In addition to providing a window to "see," these photographs can also engage readers in analyzing research findings. Figure 2.5 from a 1983 *American Scientist* article (Diener 1983, p. 483) is a case in point, which illustrates an experiment designed to test the nature of viroid, a substance believed to cause infectious diseases in plants.

In the experiment, researchers subjected multiple healthy and diseased tomato samples to electrophoresis, with the following results:

- Sample 1 (healthy tomato) revealed three bands: one for DNA and two for RNA (known as 5S RNA and transfer RNA).
- Sample 2 (diseased tomato) revealed the same bands as sample 1 and one extra: viroid.

The comparison between samples 1 and 2 thus provides physical evidence of the existence of viroid in diseased tomatoes.

- Sample 3 (diseased tomato with an added enzyme that digests DNA) revealed the same viroid band and RNA bands, but no DNA band.

Because viroid, as the other two kinds of RNA, was not digested by the added enzyme as the DNA band was, this is evidence that viroid is not DNA and may be a type of RNA.

- Finally, sample 4 (diseased tomato with an added enzyme that digests RNA) resulted only in the DNA band.

Because the added enzyme digests RNA, the disappearance of the viroid band, together with other RNA bands, confirms that viroid is a type of RNA.

Figure 2.5 thus functions as a scaffolding tool, and public readers are not excluded from analyzing the data just because they have no prior training in electrophoresis or do not have formal knowledge of, say, 5S RNA or viroid RNA. Rather than being the party who is "deficient" in background knowledge, readers are made capable of evaluating research findings. Though such experiences do not begin to reflect the full range of public engagement in science, they contribute to the broad agenda for public dialogue, at least at the normative level (see Jackson et al. 2005).

Despite these values of electrophoresis photographs for popular science communication, not all electrophoresis results are captured this way, especially in more recent publications. Consider Fig. 2.6, which is part of a larger illustration that appeared in a study of P-glycoprotein, a cell membrane protein believed to cause drug resistance. In parts not reproduced here, the illustration shows sample cells being sheared and membrane proteins being isolated. The membrane proteins are then subjected to electrophoresis, with the result shown in Fig. 2.6. Column *a* represents testing results for drug-sensitive cells, and column *b* represents those for drug-resistant ones. The horizontal lines drawn across each column are their respective bands. The two columns, as Fig. 2.6 shows, have identical bands except for an extra one in column *b*, which is colored green and marked by an arrow. A specific agent (an antibody) is then added; it binds with the molecule in that extra band and reveals it as P-glycoprotein.

To capture electrophoresis, as Fig. 2.6 does, in illustrations rather than photographs is an interesting rhetorical choice. Doing so allows a visual creator to integrate this procedure into larger illustrations of complex experiment processes. It also, arguably, helps distill complex photographic details so that a non-expert audience may more easily identify elements of interest. This, as discussed earlier, is the reason for favoring interpreted images over mechanical images. But photographs of electrophoresis, I

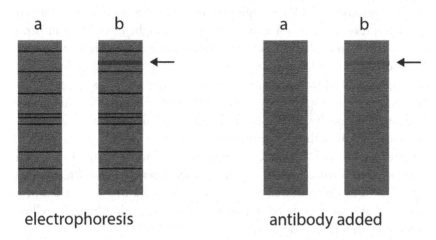

electrophoresis   antibody added

Fig. 2.6 Illustrated electrophoresis reveals P-glycoprotein. Redrawn based on Kartner and Ling (1989, p. 48) (color figure online)

argue, are different from other kinds of photographic evidence. While an X-ray photograph of the human skull would contain too many visual details that obscure the object of interest, say a lesion (Galison 1998), in the case of electrophoresis, the photograph contains virtually no visual excess. All that is photographed *is* the object of interest: the number, length, and position of the bands. When we transform these photographic details to illustrations, we are not so much providing more accessible visual evidence as reducing that evidence. A more vigilant viewer may also see this as a slippery road toward "cleaning up" primary evidence. Certainly, I am not suggesting that photographic representations of electrophoresis are free from selection or manipulation. As Knorr-Cetin and Amann (1990) reported, with photographs, researchers can perform a range of acts to make certain bands disappear, including cutting off bands, reducing film exposure time, or turning off the electrophoresis apparatus at a certain time. But if that is so, imagine the "artistic leeway" built into the illustration. It is no coincidence that illustrations of electrophoresis would never be accepted as publishable evidence by field journals.

Of course, what counts as valid evidence in field journals is a complex topic (see Frow 2012) and beside the point here. For our focus on public communication, I argue that illustrative representations of electrophoresis are inadequate for conveying the full range of implication and complication of DNA fingerprinting. From a superficial stance of visual impression, it is not easy to imagine straight lines in an illustration as traces of biological samples. Indeed, the defined lines may give the erroneous impression that genetic samples become visible lines after electrophoresis. More importantly, an illustrative view falsely represents the nature of electrophoresis: as objective "facts" that defy questions and judgment, especially questions and judgment from non-expert audiences unfamiliar with the technique.

Consider Fig. 2.7, which shows how electrophoresis is used to match suspect DNA with crime scene evidence in forensic DNA fingerprinting. Clear and uniform bands run down four columns, and it is but a quick visual scan for one to conclude that suspect 2 (outlined in yellow "matching" frame) has done it.

In reality, however, the bands formed during electrophoresis are much fuzzier (as seen in Figs. 2.4, 2.5). And interpreting these bands

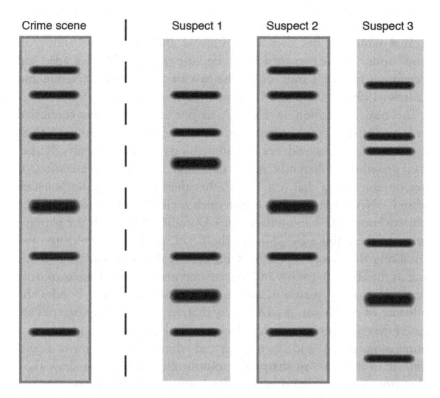

Fig. 2.7   DNA fingerprinting matches suspect DNA with crime scene evidence (Leja 2010) (color figure online)

is far from a clear-cut, straightforward process—hence the controversy of using DNA evidence in court. As Kirby (1992) wrote, "The ability to declare a match between two profiles...can be influenced by many factors"; one important factor is band shifting, which "is the phenomenon where DNA fragments in one lane of an electrophoresis gel migrate across the gel more rapidly than *identical* fragments in a second lane" (p. 119, emphasis mine). Various reasons may cause band shifting, such as different quantities of DNA being added in different lanes or gradient pockets in the gel (Kirby 1992, p. 123). An observer, therefore, has to determine whether the discrepancies in the bands are a result of acceptable band shifts or a non-match. Gray areas also set in in the cases of "extra bands." When sample A has identical bands as sample B plus some extras, it is possible to conclude that (1) the two are a match because sample A has been contaminated and therefore is taking on extra bands, (2) the two are a match because sample B has degraded and therefore is missing bands, or (3) the two are a non-match (Nelkin and Andrews 1999; Neufeld and Colman 1990).

The neat illustration in Fig. 2.7, in one decisive move, eradicated these potential complications and "black-boxed," as Latour (1998) would say, the visual evidence into something that resists inquiry. Rather than presenting inherently messy data that call for interpretation, it presents something that just "is." Rather than illustrating the "science" behind DNA fingerprinting, it disregards social debate surrounding the subject (Long 2007) and reflects what Durodié (2003) calls the patronization of the publics by "diluting the detail, eroding the evidence and trivialising the theory" (p. 85). Reinforced is the positivist view of science as the absolute reality, and lost are opportunities for readers to truly access data, raise questions, and consider issues of trust. Besides the influence of positivism, it seems likely that, in this case, photographs are passed over also because of their lack of visual glamor. In popular communication, grainy and often black-and-white bands are no competition, or so it seems, to sharp and colorful illustrations that draw their ethos from sophisticated design software and technical precision.

Symbolic Photographs: Visual Catachresis

Electrophoresis bands may lack visual appeal, but that does not mean they are affect-less. As rich semiotic objects, photographs are adept at evoking viewer reactions and emotions—just consider the everyday pictures we take of ourselves and our surroundings. Photographs used in the popular communication of genetics are no exception. The crossbreeding photographs from the early twentieth century signify the depicted plants and animals as experiment specimens, not part of nature to take delight in. The animals look away rather than at the viewer, forming what Kress and Leeuwen (2006) called "offer images": that is, what is photographed is offered "to the viewer as items of information, objects of contemplation, impersonally" (p. 119). The electrophoresis photographs, in their sober depiction of unadorned bands, convey much the same feeling and impression.

If this sense of detached objectivity or contemplation is often expected in scientific visuals, photographs that appeared in the last 20 or so years in the popular communication of genetics are charting new territories. Following Bloomfield and Doolin (2012), I call these images symbolic photographs, photographs that are less interested in being scientific evidence than creating or repurposing signs to channel and foster viewer reactions and emotions. An example, in the extreme, should demonstrate what I mean.

In 2003, to protest against transgenic cattle programs and the lift of a moratorium on genetically modified organisms (GMOs), the New Zealand activist group Mothers Against Genetic Engineering in Food and the Environment (MAdGE) erected a billboard at sites in Auckland and Wellington (Bloomfield and Doolin 2012). Pictured on the billboard is "a photographic image of a naked, four-breasted young woman, kneeling on all fours in side profile, with her breasts hooked up to a dairy milking apparatus and a red 'GE' brand on her buttock. The accompanying press release was titled 'Why Not Just Genetically Engineer Women for Milk?'" (Bloomfield and Doolin 2012, p. 515). Clearly, this billboard has other purposes than demonstrating the process of genetic engineering or otherwise serving as scientific evidence.

Such extreme examples, as may be expected, are not found in mainstream popular science publications or mass media. But similar visual and semiotic gestures exist, enabled, in no small part, by widely available graphic editing software such as Photoshop. An article on artificial human reproduction (Reynolds 2005, p. 72), for instance, presented a photographic image of a human baby, umbilical cord attached, inside a transparent bubble. The bubble, as the article suggested, stands for the artificial amniotic sac or more generally, the artificial womb. From outside of the picture frame, a hand reaches down and caresses the bubble.

The image features a calm blue color, the hand looks markedly female with a gentle gesture, and the baby inside the bubble looks comfortable and healthy. But the photograph, as a whole, looks eerie. The grown baby looks naturally human, but floating in a transparent bubble, it is apparently not a natural baby. The hand from above creates the same effect: it looks to be the hand of a mother, but common sense tells us that no natural mother can touch her baby thusly. Although fetus-in-womb drawings date back to as early as Leonardo Da Vinci, the example here has a distinct visual impact by being a photograph, a visual genre that is conventionally associated with recorded reality. Blending the perceived "real" and the apparently "unreal," the photograph is capable of creating the state of the uncanny (Mori et al. 2012) or, at least, an uncomfortable feeling of someone trying to play God.

The most common symbolic photographs used in popular science and media reports of genetics are, as in the cowgirl example, about GMOs. For example, an ear of corn would have its husk peeled back to reveal not yellow kernels but colorful pills—implying the genetic engineering of corn to produce proteins and other elements of medicinal value (Ferber 2003). Succulent vegetables and fruits would take a shot of suspect fluids from a syringe—a most popular portrayal of GMOs (see Fig. 2.8). Apples would be patched up with differently colored parts to suggest genetically engineered apples with extra apple genes (Pollack 2012), or they would slice open to reveal the inside of an orange, implying, ostensibly, a more audacious transgenic attempt. These examples, as may be imagined, generally accompany anti-GMO discussions. Though not nearly as extreme as the cowgirl example, they are visually shocking by fusing the real and the fantastic and by portraying the socially and culturally "unnatural."

While the semantic gist of these Frankenstein food photographs is easy enough to grasp, their semiotic effect on publics' perception of genetics is far from simple. A useful lens to examine that effect is catachresis, which, according to classic Quintilian definition, is a figure of speech wherein "the nearest available term" is adapted "to describe something for which no actual [i.e., proper] term exists" (Parker 1990, p. 60). This definition, I propose, can be productively extended from verbal adaptation to photographic adaptation. For instance, in artificial human reproduction, the process of implanting embryos on endometrial cells cannot be readily captured; similarly, the molecular-level differences between GMOs and non-GMOs are not visible to the naked eyes. There is, in other words, no actual way to photograph these scenarios for a viewer. In this vacuum, the nearest available means of representation—or more precisely, what a visual creator considers the nearest

Fig. 2.8 Catachrestic photograph of genetically modified tomato (color figure online)

available means—is used. Thus, a transparent bubble stood in for an artificial womb, female breasts represented "milk" genes, colored pills substituted for protein genes, and physical injections of fluid replaced genetic engineering.

What is remarkable about these substitutions is their ambiguous intention and effect. On the one hand, it is possible to see them as metaphorical expressions, clever figures of speech wherein a visual creator uses a familiar term to elucidate an unfamiliar or abstract concept. On the other hand, it is possible to see them as a form of misuse, taken after the original meaning of *catachresis*, from the Greek *katakhresis*. How to delineate these two different interpretations of catachresis has troubled rhetors for centuries. For Quintilian, catachresis is not a metaphor: To use metaphor is to replace a proper term with one transferred from another place; to use catachresis is to adapt a word for a situation where no proper term exists (Parker 1990). However, this rationale easily breaks down. As Quintilian himself admitted, one may "indulge in the abuse of words even in cases where proper terms do exist" (Parker 1990, p. 61). Later rhetors, attempting to clarify the matter, came to define catachresis as a "bad" metaphor. Cicero defined it as "an abuse of metaphor, the wrong or inexact use of it as a substitution for the proper term," and Northrop Frye called it the "unexpected or violent metaphor" (Parker 1990, p. 61). For these scholars, a metaphor is the friendly and neighborly borrowing of words, whereas catachresis is constrained and forced (Parker 1990).

To visualize "freedom" in the form of a flying eagle, many (at least in the U.S. context) will agree is friendly; to visualize women with cow-like breasts, most readers would concur is unexpected and forced. At the very least, such representations are semiotically excess; that is, what is depicted in the photograph is "greater than [what] the material foundation can warrant" (Anderson 1996, p. 55). This excess is what made the cowgirl billboard backfire: the image was critiqued by many as offensive and erroneous and ultimately ruled by the New Zealand Advertising Standards Complaints Board as a depiction that "[distorts] the debate on genetic engineering" (Bloomfiled and Doolin 2012, p. 519). The Frankenstein food catachrestic photographs, because they are less provocative, receive no such backfire, but from a semiotic perspective, they are similarly excessive. Consider the common approach to cultivating GMOs: it starts with splicing a target gene or genes into harmless bacteria, which then slip the gene(s) into a plant cell on contact; after taking up the gene(s), the cell divides and multiplies to generate a seedling, which then bears edible vegetables or fruits. This and other such processes, then, bear little resemblance to injecting fluids into grown and harvested vegetables or fruits, as Fig. 2.8 would have us believe.

But from an activist stance, a catachresis' ability to misuse, to shock, to insinuate subversive voice into the dominant discourse is its precise attraction. Termed a "secondary original" by Jacques Derrida, "catachresis is both an impropriety and an opportunity" because it is when we disregard the apparently proper meaning of a sign that we are liberated to expose a reconfigured relation to it (Hawthorne and Klinken 2013, p. 160). This makes catachresis an attractive frame of reference used by, among others, postcolonial and gender scholars to challenge the status quo (Hawthorne and Klinken 2013).

For the same reason, catachrestic photographs can function as subversive voice to formal scientific institutions' (and their media allies') portrayal of genetics as that which is scientifically groundbreaking and socially beneficial in an absolute sense. Catachrestic photographs do so by appealing to the powerful concept of nature as "good, pure..., balanced and harmonious..., a self-balancing system, a force that is best left to its own devices, a system which will continuously 'sort itself out'" (Hansen 2006, p. 813). Set on this background, genetic manipulation becomes unnatural, "a contaminating and polluting interference," and "has unpredictable outcomes" (Hansen 2006, p. 815). Employing concrete and incongruent images, catachrestic photographs make these abstract ideological stances instantly recognizable and relatable to a wide public audience. While the images may not be "real," they are the "paramount reality" people believe to be real and embody people's local understanding, value, and practice—in short, their "common sense" (Bauer and Gaskell 2008, p. 345).

Conceived in these terms, catachrestic photographs are commendable and valuable in creating a postmodern landscape of public science communication, a landscape that celebrates multiple agents, viewpoints, and voices. Created for and possibly by publics, they make tangible the abstract, unfamiliar, and threatening idea of genetic modification in ways that mount a creative resistance to it (Bauer and Gaskell 1999, 2008). They help to express public concerns; provide entry points for reflection; conjure up images that bridge science and society; and reveal, in a most relentless way, the broad and potential social and ethical implications of genetic research.

Just as catachrestic photographs mount a creative resistance to formal science, "pro-GMO" symbolic photographs constitute a resistance to public objection. Ironically enough, these photographs appeal to the same powerful concept of "nature," only in an opposite way, that is, by showing the apparent "naturalness" of GMOs. In these instances, GM papayas look just like ordinary papayas, GM corn grows like regular corn, and GM cattle appear no different than ordinary cattle (see, e.g., Harmon 2014; U.S. Food and Drug Administration 2014). Elsewhere, other scholars have made similar observations: Finland's transgenic medicine cow Morrow would be photographed in her cowshed, looking

like any other cow (Väliverronen 2004), and a piglet genetically modified to aid transplant research would be photographed next to a regular
piglet and look equally cute (Mellor 2009). In all of these instances, we
witness apparently real evidence of GMOs being familiar, comforting,
and pleasant, the same way nature intends them to be.

The contrast between the catachrestic and the pro-GMO photographs reflects the alternative experiences of different stakeholder groups
in the GMO debate. Science communication per se cannot hope to
change these alternative experiences, which are bound up with group-
specific values, practices, and inter-group communications. For either
party, the different representations of the "other" present a challenge,
but the "other" and its alternative representations, as Bauer and Gaskell
(1999) argued, must at the very least be acknowledged. Within the
postmodern landscape, the "other" is "not necessarily a problem for the
subject, but may help to structure and stabilize the subject's experience
of the world within a constant flux of events.... It is through the contrast of divergent perspectives that we become aware of representations,
particularly when the contrast challenges our presumed reality and is
resisted" (Bauer and Gaskell 1999, p. 169). As such, the question about
communication, to borrow Bauer and Gaskell's (1999) words, is not
how to reach consensus but how to coordinate conversations between
different voices.

With the photographic representations of GMOs, such coordina-
tion needs to consider at least two factors. First, as mentioned earlier,
both anti- and pro-GMO symbolic photographs base their appeal on
"nature" as that which is "universal," "right," "eternal," and "non-
negotiable" (Hansen 2006, p. 813). But precisely because "nature" is
presumed to be universal and non-negotiable, these images, in their
respective ways, resist questions and debate and create what Hansen
(2006) terms a "discursive stopper." They both assume certain "boundaries that separate the natural from the non-natural" without questioning
those boundaries (Hansen 2006, p. 830). But clearly, where those
boundaries lie is different for different people: for some, being natural
means being free of any and all genetic alteration; for others, it means
being intrinsically the same notwithstanding genetic alteration. Without

both sides questioning their definitions of boundaries, the "natural" or "unnatural" photographs do little to move forward productive dialogue.

Second, when catachrestic photographs are repeatedly used in mass media (as they already are), they cease to be figures of speech but dead metaphors whose authority is taken for granted or at least evades conscious judgment. They become what Marilyn Strathern calls habitual images that "shape cultural expectations and the emotional structure of everyday beliefs" (Nelkin and Lindee 2004, p. 12). As habitual images, they affirm group values and facilitate "shorthand," taken-for-granted inter-group communication but do not facilitate dialogue with external groups who do not share those values. As shorthand communication, they also discourage abstract arguments on either side of the GMO debate. On the pro-side, researchers argue that "transgenic DNA does not differ intrinsically or physically from any other DNA already present in foods" and that the movement of genes from one population to another is a natural phenomenon that exists independently of genetic engineering (Nicolia et al. 2014, pp. 3, 6). On the anti-side, critics argue that GM crops can pose ecological risks because their effects are not locally contained: pest-resistant GM crops could release their insecticidal compounds into the soil through root and into the air through pollens, impacting non-target organisms and insects. These finer-grained arguments are what will enable publics to investigate, question, and interrogate GMO research, and neither the catachrestic or pro-GMO photographs are particularly adept at facilitating such conversations.

Symbolic Photographs: The Human Drama

Aside from catachrestic treatments, another group of symbolic photographs is prevalent in the popular communication of genetics. Their symbolism centers around the human body, what Douglas (1970) called a natural and prevailing symbol in human interaction and communication. The use of this symbol, as Douglas (1970) argued, is subject to social constraints, which determine what kinds of bodily representations are acceptable; at the same time, the representations of the body also serve to maintain or recreate those constraints. In the case of the cowgirl billboard, social outrage erupted precisely because the billboard digressed from, and attempted to drastically recreate, what is considered acceptable ways to represent the (female) body.

Most human bodies portrayed in popular science publications and the mass media are far less controversial, though they are equally significant in their configuration (and reconfiguration) of the public perception of modern genetics. These photographs cast the human bodies in different semiotic roles that, together, stage what van Dijck (1998) would call the human drama of genetics. Centered around medical urgencies and human needs, this drama is what pushes genetics into the foreground of public recognition and political attention (Dijck 1998).

The Victims

Any good drama needs a victim to be saved (or has, alas, been lost). In the human drama of genetics, the victim is the unwell, the potentially unwell, or more broadly, the "target" of genetic research and medical intervention. Earlier photographs in this category tend to focus on the apparently ailing: for example, newborn babies with physical defects or adults with observable medical symptoms. The contemporary cast of "victims," however, is significantly broader. In addition to those who are unwell in the traditional sense, we see photographs of the elderly (e.g., Fig. 2.9), the physically unfit (e.g., overweight), the socially maladjusted (e.g., alcoholics), and the statistically vulnerable (e.g., Google co-founder Sergey Brin, who was reported to have higher risks for Parkinson's in a *New York Times* article [Helft 2008]).

In addition to humans, anthropomorphized animals and plants occasionally play the role of the victim, for example, cows contracted with mad cow disease or plants that face epidemics. In these moments, animals and plants are photographed very differently from the way they were in the early twentieth-century informative photographs. Rather than being experimental specimens captured in "offer images" for impersonal contemplation (Kress and Leeuwen 2006), they are individual entities with human-like characteristics. Animals are by far

Fig. 2.9 Nursing facility-bound elderly woman awaits genetic solution. Because of copyright restrictions, the original photograph used in Rusting (1992, pp. 130–131) cannot be reproduced. The photograph used here resembles the original in essence and style. Credit/Copyright Attribution: Alexander Raths/ Shutterstock (color figure online)

the easiest to humanize: In close-up shots, their facial expression easily resembles that of humans and evokes human emotions. Plants are less conducive to anthropomorphism, though not impossible. In "Can this fruit be saved" (Koeppel 2005), we learn the impending pandemic faced by the banana plant due to its lack of genetic diversity. To demonstrate bioengineering as a potential savior, a photograph depicts a banana in a surgical tray and being operated on. Its peel is carefully removed by a pair of tweezers, as if a layer of skin removed from a human patient.

Photographs as these do more than providing a neutral context for genetic research. At the most basic level, they invite empathy from readers toward the photographed fellow humans or human-like life forms. A sense of personal identification is also palpable: I, my families, or those around me could be one of these people. Portraying real victims awaiting, the photographs also create a sense of social urgency. The photograph of a young girl, a cystic fibrosis patient, breathing through the nebulizer (Wright 1999) constitutes a powerful plea to loosen the human embryo research restriction so genetic therapy may be invented to "knock out" defect genes in unborn embryos. These photographs, then, help to invoke public support for genetics, for funding, and for favorable research policies—at least within the narrow confines of medical intervention.

Aside from scientific and social motivations, our apparent willingness and readiness to put human bodies—or human-like bodies—on display as "victims" reflects and reinforces the society's acceptance of and trust in genetics as the ultimate medical solution. This promised solution, however, also normalized our bodies in new and potentially problematic ways. While a 1980s study that probes the genetic predisposition to alcoholism opted to use a cartoon illustration of two gentlemen drinking in what seems a social context (Bower 1988), recent studies have moved on to photographing, in close-up shots, solitary, secretive, and desperate drinkers (e.g., Blum et al. 1996; Nurnberger and Bierut 2007).

This new norm, when it is—as it seems already—accepted by the publics, aligns with the reductive way of thinking about genetics, a mentality that Nelkin and Lindee (2004) termed genetic essentialism and Rose (1995) named neurogenetic determinism. This mentality reduces humans to their molecular entity, thus equating human beings, notwithstanding their social, moral, and behavioral complexities, to their genetic makeup (Nelkin and Lindee 2004). It attempts to find *the* cause for complex and dynamic biological and social phenomena (e.g., depression, intelligence, aging, or violence) in one or a few genes and, when appropriate, find ways to modify the genes to relieve personal pains and social disorders (Rose 1995).

Among the scientific community or at least as conveyed in peer-reviewed publications, such a mentality is not endorsed. Most researchers recognize that no single gene or genetic event underlies simple, let alone complex, life forms and functions. In depression studies, for example, despite decades of efforts, "no single genetic variation has been identified to increase the risk of depression substantially" and "multiple genetic factors in conjunction with environmental factors" are believed to play a role (Lohoff 2010, p. 539). Similarly, in the emerging study of pathological gambling, researchers acknowledge that family environment, traumatic life events, and social factors (such as the availability of legal gambling) all factor into the development and persistence of pathological gambling (Lobo and Kennedy 2009). When genetic influences are identified, researchers are careful to acknowledge limited research methods and use languages such as "contribute," "associate," and "explain" rather than the absolute "cause" or "determine."

These understandings, however, have not deterred media reports from finding the depression gene, gambling gene, fat gene, financial debt gene, et cetera. It is hard to say that media outlets are solely responsible for such sensational reporting; more likely, it is caused by the combined pressure of reporters, editors, and scientists to promote their respective research and communication products (Petersen 2001; Treise and Weigold 2002). Each report helps us to get a little closer to genetic essentialism and determinism, with far-reaching scientific and social ramifications: the pursuit of a linear and simplistic genetic research paradigm, a society absolved from trying to find social solutions for social problems, individuals absolved from taking responsibilities for their actions, and scarce public resources being wasted on ill-defined research (Rose 1995; Nelkin and Lindee 2004).

The Heroes

When there are victims, there must also be heroes. In the human drama of genetics, the hero, as may be expected, is played by scientists, researchers, and medical doctors, who attempt to find and/or implement genetic solutions to solve human suffering. This depiction of scientists as intellectual heroes is an age-old scheme (see Cartwright 2007) and also reflected in verbal reports of genetics (Petersen 2001), but carried into photographic forms, it is capable of evoking even stronger viewer reactions and emotions.

According to van Dijck (1998), "Until the 1970s, pictures of scientists were still rare in popular stories. If photographed at all, scientists were usually represented as part of a group in the non-descriptive, passport-size pictures that illustrated the articles" (p. 18). van Dijck (1998) theorized that the rise of scientist photographs in the popular communications of genetics had to do with the boom of the biotech companies in the 1980s: the photographs served to project the personal qualities and fame of the scientists onto an industry that had yet to produce physical products and thus helped to boost investor confidence.

My work, however, suggests a much earlier timeline for the use of hero photographs. They started to appear in popular science publications as early as the 1920s and 1930s, in the heyday of classical genetics. There are, however, certain differences between earlier and contemporary hero photos. The earlier photographs were often not posed; that is, the scientists were photographed engaging in their (presumably) usual work activities, conversing with each other, looking down into their microscopes, or working with experiment samples or laboratory equipment. These shots were often mid-ranged to capture the surroundings of the scientists and to give the impression of scientists at work, not of scientists per se. Even when the heroes were shot close up, they often faced an angle oblique from the viewer and did not make direct eye contact with the viewer.

Recent hero photographs are more likely to be close-up shots and posed frontal shots. Iconic displays of science and medicine (petri dishes, medications, non-descriptive specimens, and complex-looking tools), rather than being worked on by the heroes, serve as visual background or decoration. In the foreground, the portrayed characters are not engaged in any activities but look directly at the reader, often smiling (Fig. 2.10) but sometimes looking serious.

Fig. 2.10 Oral biologist Dr. Jeffrey Hillman, an advocate of transgenic microbial medicine (Sachs 2008, p. 66). Courtesy of Paul Figura (color figure online)

The different horizontal angles assumed by contemporary and earlier hero photographs are significant. As Kress and van Leeuwen (2006) pointed out, the horizontal angle "encodes whether the image-producer (and hence, willy-nilly, the viewer) is 'involved' with the represented participants or not. The frontal angle says, as it were, 'What you see here is part of our world, something we are involved with.' The oblique angle says, 'What you see here is *not* part of our world; it is *their* world, something *we* are not involved with'" (p. 136). When the hero faces the viewer with a direct gaze, the image constitutes a "demand" photograph; that is, the viewer is explicitly asked to enter into some imaginary relation with the person photographed (Kress and Leeuwen 2006). When the heroes are smiling, "the viewer is asked to enter into a relation of social affinity with them" (Kress and Leeuwen 2006, p. 118); when they look serious, the viewer is asked to trust the integrity of their work.

Aside from the economic incentive van Dijck (1998) mentioned, these photographs reflect scientists and researchers' motivation to "use the media to promote the importance of their work, to improve their public image in order to assure continuity of public funding for their research and to counter negative images of genetics" (Petersen 2001, p. 1257). While supporting data are difficult to find, given the contemporary emphasis on public engagement, it seems plausible that there is also genuine interest behind these photographs to involve the publics in the world of science or present the prospect of direct conversation. At the very least, contemporary photographs raised the stake for scientists' public accountability, which is best highlighted in the depiction of "bad" or controversial scientists. In these moments, readers question why scientists wear top-to-bottom protective suits when collecting "safe" biologically engineered bacteria (Baskin 1988). They witness the gleeful smile of South Korean cloning specialist Woo Suk Hwang, who reported fraudulent success in creating human stem cell lines (Hooper 2006). When public readers are explicitly invited to view the scientists as fellow citizens, as one of "us" and to see their work as part of social activities, as "our" work, they are "sanctioned," to an extent not possible even 20 years ago, to monitor and question science and scientists.

The Everyday Stakeholders

Besides the more dramatic roles of victims and heroes, the human drama of genetics also features everyday stakeholders: anyone who may be implicated in or impacted by genetic research—read, every one of us. Photographs of these characters often accompany reports that are beyond the narrow confines of medical urgency and as such, take a more lighthearted approach. An article (Stover 2002) that discusses the genetic relationships between species, for example, shows a chimpanzee playfully touching a human while an orangutan looks on; an article (Ast 2005) that explains the common genes shared among species photographs a human holding his rodent relative the mouse; and an article (Arking 2003) that examines the promise of genetic intervention to extend human life and vitality pictures a senior but fit citizen in a bright outfit cycling by (Fig. 2.11).

Visually interesting or relatable, these photographs serve to spark readers' general interest or personal curiosity in genetics even when there is a lack of explicit medical urgency. One, however, cannot help but notice that in photographs after photographs, the "everyday"

Fig. 2.11 Senior but active man sets the context for aging study. Because of copyright restrictions, the original photograph used in Arking (2003, p. 509) cannot be reproduced. The photograph used here resembles the original in essence and style. Credit/Copyright Attribution: Michaeljung/Shutterstock (color figure online)

human faces are those of men. The female body is conspicuously scarce, though there is no shortage of it in the victim category. These gendered visual choices, conscious or otherwise, reflect and reinforce the social perception that women are more likely to be victimized and that victimization is a feminine or potentially feminizing experience (Howard 1984). More particularly, in this context, they reflect and reinforce the historical and prevailing "male focus" in genetic research. As Clayton and Collins (2014) wrote in a recent *Nature* article, the "norm" in biological and clinical experiments is to use male animal models and cell lines, for several reasons: the belief that female animals' hormone cycles introduce complications into the experiments; a lack of understanding in the effect of sex on research findings; and the simple matter of following established protocols. However, the supposed female hormone influence is a questionable basis, as female mice may exhibit no more influence by their hormone cycles than male mice do; more significantly, researchers have started to realize that sex is a significant correlating factor in biological and clinical studies, influencing, among other things, subjects' reaction to treatments (Clayton and Collins 2014). With these understandings, the National Institutes of Health is taking measures stronger than its previous perfunctory expectation and enforcing gender diversity by requiring grant applications to balance their use of male and female samples and by developing training modules to help investigators evaluate sex differences (Clayton and Collins 2014).

As the research paradigm shifts, we may hope to see more diverse media representations of the everyday stakeholders and reconstruct our perception of females and their stakes in genetic research. This need for change is not limited to gender either, for not coincidentally, most of the male stakeholders currently pictured are also white. And probably not surprisingly, the same scientific backstory holds true. The majority of current genetic studies and verified data involve participants of European descents. As Haga (2010) reported, "The initial study populations of 79% of the US GWAS[2] publications were all white; 75% of the replication sample populations were also all white.... Overall, 92% of US GWAS participants were white, followed by African-Americans (3%)" (p. 81). Established conventions and datasets are, once again, cited as reasons for such practice, as is the difficulty in recruiting minority participants (Haga 2010; Knerr et al. 2011).

As a result, current research findings and clinical guidelines cannot be generalized to the wider population (Knerr et al. 2011). In the case of genetic testing, for example, only limited testing is currently available to minority populations. Of the three major companies that offered direct consumer genetic testing, two of them (23andMe and deCODEme) indicated that 16 and 11 of the 22 diseases they tested were not applicable to individuals of non-European ancestry (Haga 2010). The third company (Navigenics) chose not to reveal this limitation in their test descriptions, but their consent documents and test reports indicated that "most of their testing is based on studies of people of European ancestry and, therefore, [they] are uncertain as to whether the results are applicable to people of other backgrounds" (Haga 2010, pp. 81–82). If this healthcare inequality strikes one as astonishing, it is only going to increase with time and continuous research that follows the well-trodden path. The social and political ramifications of such practice are too obvious to need belaboring here.

When today's readers turn to Charles Darwin's *On the Origin of Species*, they are stricken by the conspicuous use of "man" as the representation of the human race: "nature gives successive variations; man adds them up in certain directions useful to him" (p. 35). The contemporary society has, fortunately, become more conscious in adopting non-biased language. The same, I argue, is imperative in visual representations and, more fundamentally, in research practices. For a field that has a professed goal in improving human (read, all human) welfare, we cannot afford anything less.

Conclusion

Even though I separated informative photographs that serve as visual evidence from symbolic photographs that evoke human emotions, this separation is more of a necessity in linear discussion than a suggestion of clear demarcation. Even when a photograph's primary purpose is to present cognitive information, it still embodies affective elements; similarly, even as a photograph's ostensible purpose is to create emotional appeal, it can belie scientific backstories.

Photographs, as this chapter shows, play multiple and sometimes conflicting roles in the popular communication of genetics. They can be used to convey visual evidence, to invite scrutiny, to attract and engage, to insert subversive voice, and to lobby for public support, all of which help to reflect as well as shape publics' evolving perception of genetics. Because of these varying possibilities, the use of photographs is not a straightforward process. This semiotically rich visual genre can be used (or avoided) to frame genetic research and public discourse in different ideological context with alternative value stances, sometimes in subtle ways that resist questioning, unless we pay conscious attention, as this chapter attempted to do. If one thing is certain, it is that photographs will continue to be prevalent in the popular communication of genetics. Their effects and implications are thus important to consider for everyone involved in creating these images, from scientists, science communicators, visual designers, activist groups, to public members.

Notes

1. The determination of the DNA structure relied on key information provided by X-ray crystallography, a diffraction imaging technique pursued by Maurice Wilkins and Rosalind Franklin at King's College, London. Most specifically, one such image, taken by Rosalind and her student Raymond Gosling, allowed James Watson and Francis Crick to develop their helical DNA model. This image, though often nicknamed "Photo 51," is not the kind of conventional photograph discussed here.
2. GWAS stands for genome-wide association study, which examines the complete sets of DNA from many people to find genetic markers associated with certain diseases.

References

Anderson, J. A. (1996). *Communication theory: Epistemological foundations*. New York: The Guilford Press.

Arking, R. (2003, November–December). Aging: A biological perspective. *American Scientist, 508*–515.

Ast, G. (2005, April). The alternative genome. *Scientific American,* 58–65.

Baskin, Y. (1988, July–August). Genetically engineered microbes: The nation is not ready. *American Scientist,* 338–340.

Bauer, M. W., & Gaskell, G. (1999). Towards a paradigm for research on social representations. *Journal for the Theory of Social Behaviour, 29*(2), 163–186. doi:10.1111/1468-5914.00096.

Bauer, M. W., & Gaskell, G. (2008). Social representations theory: A progressive research programme for social psychology. *Journal for the Theory of Social Behaviour, 38*(4), 335–353. doi:10.1111/j.1468-5914.2008.00374.x.

Bloomfield, B. P., & Doolin, B. (2012). Symbolic communication in public protest over genetic modification: Visual rhetoric, symbolic excess, and social mores. *Science Communication, 35*(4), 502–527.

Blum, K., Cull, J. G., Braverman, E. R., & Comings, D. E. (1996, March–April). Reward deficiency syndrome. *American Scientist,* 132–145.

Bower, B. (1988, July 30). Alcoholism's elusive genes. *Science News,* 74–75, 79.

Brown, K. (2001, April). Seeds of concern. *Scientific American,* 52–57.

Brownlee, C. (2006, June 24). Nurture takes the spotlight. *Science News, 169*(25), 392–393, 396.

Cartwright, J. (2007). Science and literature: Towards a conceptual framework. *Science & Education, 16*(2), 115–139. doi:10.1007/s11191-005-4702-9.

Castle, W. E. (1905, July). Recent discoveries in heredity and their bearing on animal breeding. *Popular Science,* 193–208.

Clayton, J. A., & Collins, F. S. (2014). Policy: NIH to balance sex in cell and animal studies. *Nature, 509,* 282–283.

Cohen, F. (1994). *The scientific revolution: A historiographical inquiry.* Chicago: University of Chicago Press.

Diener, T. O. (1983, September–October). The viroid–A subviral pathogen. *American Scientist,* 481–489.

Diener, T. (2003). Discovering viroids–A personal perspective. *Nature Reviews Microbiology, 1*(1), 75–80.

Dobrin, S. I., & Morey, S. (2009). *Ecosee: Image, rhetoric, nature.* Albany, NY: SUNY Press.

Dolinoy, D. C., Weidman, J. R., Waterland, R. A., & Jirtle, R. L. (2006). Maternal genistein alters coat color and protects Avy mouse offspring from obesity by modifying the fetal epigenome. *Environmental Health Perspectives, 114*(4), 567–572.

Douglas, M. (1970). *Natural symbols: Explorations in cosmology.* London: Barrie and Rockliff.

Durodié, B. (2003). Limitations of public dialogue in science and the rise of new 'experts'. *Critical Review of International Social and Political Philosophy, 6*(4), 82–92.

East, E. M. (1910, October). The role of hybridization in plant breeding. *Popular Science, 342–255.*

Ferber, D. (2003, April). Something funny down on the pharm. *Popular Science, 78–84.*

Frow, E. K. (2012). Drawing a line: Setting guidelines for digital image processing in scientific journal articles. *Social Studies of Science, 42*(3), 369–392.

Fulton, T. (1998). *The United States senate committee on agriculture, nutrition, and forestry 1825–1998.* (No. Y 1.1/3:105–24). Washington, DC: U.S. Government Publishing Office.

Galison, P. (1998). Judgment against objectivity. In C. A. Jones, P. Galison, & A. E. Slaton (Eds.), *Picturing science, producing art* (pp. 327–359). New York: Routledge.

Haga, S. B. (2010). Impact of limited population diversity of genome-wide association studies. *Genetics in Medicine: Official Journal of the American College of Medical Genetics, 12*(2), 81–84. doi:10.1097/GIM.0b013e3181ca2bbf.

Hansen, A. (2006). Tampering with nature: 'Nature' and the 'natural' in media coverage of genetics and biotechnology. *Media, Culture and Society, 28*(6), 811–834.

Harmon, A. (2014). A lonely quest for facts on genetically modified crops. *The New York Times.* Retrieved April 23, 2015, from http://www.nytimes.com/2014/01/05/us/on-hawaii-a-lonely-quest-for-facts-about-gmos.html?_r=0.

Hawthorne, S. M., & van Klinken, A. S. (2013). Introduction catachresis: Religion, gender, and postcoloniality. *Religion and Gender, 3*(2), 159–167.

Helft, M. (2008, September 18). Google co-founder has genetic code linked to Parkinson's. *The New York Times.* Retrieved January 31, 2015, from http://www.nytimes.com/2008/09/19/technology/19google.html.

Hooper, J. (2006, August). Mr. hard cell. *Popular Science, 64–69, 89–91.*

Howard, J. A. (1984). The "normal" victim: The effects of gender stereotypes on reactions to victims. *Social Psychology Quarterly, 47*(3), 270–281.

Hunter, N. (1951, July 7). New kind of green gold. *Science News Letter,* 10–11.

Jackson, R., Barbagallo, F., & Haste, H. (2005). Strengths of public dialogue on science-related issues. *Critical Review of International Social and Political Philosophy, 8*(3), 349–358. doi:10.1080/13698230500187227.

Jones, D. F. (1919, September 6). Hybrid vigor and its meaning. *Scientific American,* 230–241.

Kartner, N., & Ling, V. (1989, March). Multidrug resistance in cancer. *Scientific American,* 44–51.

Kirby, L. T. (1992). *DNA fingerprinting: An introduction.* New York: Oxford University Press.

Knerr, S., Wayman, D., & Bonham, V. L. (2011). Inclusion of racial and ethnic minorities in genetic research: Advance the spirit by changing the rules? *The Journal of Law, Medicine & Ethics: A Journal of the American Society of Law, Medicine & Ethics, 39*(3), 502–512.

Knorr-Cetina, K., & Amann, K. (1990). Image dissection in natural scientific inquiry. *Science, Technology and Human Values, 15*(3), 259–283.

Koeppel, D. (2005, August). Can this fruit be saved? *Popular Science,* 60–67, 104–105.

Kress, G. R., & Van Leeuwen, T. (2006). *Reading images: The grammar of visual design* (2nd ed.). London: Routledge.

Latour, B. (1998). How to be iconophilic in art, science, and religion. In C. A. Jones, P. Galison, & A. E. Slaton (Eds.), *Picturing science, producing art* (pp. 418–440). New York: Routledge.

Leja, D. (2010). National Human Genome Research Institute. DNA Fingerprinting. Retrieved April 4, 2015, from http://www.genome.gov/dmd/img.cfm?node=Photos/Graphics&id=85150.

Lobo, D. S., & Kennedy, J. L. (2009). Genetic aspects of pathological gambling: A complex disorder with shared genetic vulnerabilities. *Addiction (Abingdon, England), 104*(9), 1454–1465. doi:10.1111/j.1360-0443.2009.02671.x.

Lohoff, F. W. (2010). Overview of the genetics of major depressive disorder. *Current Psychiatry Reports, 12*(6), 539–546. doi:10.1007/s11920-010-0150-6.

Long, H. (2007). DNA profiling: The ability to predict an image from a DNA profile. In H. Coyle (Ed.), *Nonhuman DNA typing: Theory and casework applications* (pp. 185–203). Boca Raton, FL: CRC Press.

Loughry, W. J., Prodöhl, P. A., McDonough, C. M., & Avise, J.C. (1998, May–June). Polyembryony in armadillos. *American Scientist,* 274–279.

Mellor, F. (2009). Image–music–text of popular science. In R. Holliman, E. Whitelegg, E. Scanlon, S. Smidt, & J. Thomas (Eds.), *Investigating science communication in the information age* (pp. 205–220). Oxford: Oxford University Press.

Meyer, E. T. (2007). *Socio-technical perspectives on digital photography: Scientific digital photography use by marine mammal researchers.* Ph.D. Dissertation. Available from ProQuest Dissertations & Theses Full Text (304857203).

Millius, S. (2011, July 30). DNA hints at polar bears' Irish ancestry. *Science News,* 5–6.

Mori, M., MacDorman, K. F., & Kageki, N. (2012). The uncanny valley. *Robotics & Automation Magazine, IEEE, 19*(2), 98–100. doi:10.1109/MRA.2012.2192811.

Nelkin, D., & Andrews, L. (1999). DNA identification and surveillance creep. *Sociology of Health & Illness, 21*(5), 689–706.

Nelkin, D., & Lindee, M. S. (2004). *The DNA mystique: The gene as a cultural icon.* Ann Arbor, MI: University of Michigan Press.

Neufeld, P. J., & Colman, N. (1990, May). When science takes the witness stand. *Scientific American,* 46–53.

Nicolia, A., Manzo, A., Veronesi, F., & Rosellini, D. (2014). An overview of the last 10 years of genetically engineered crop safety research. *Critical Reviews in Biotechnology, 34*(1), 77–88.

Nurnberger, J. I., & Bierut, L. J. (2007, April). Seeking the connections: Alcoholism and our genes. *Scientific American,* 46–53.

Parker, P. (1990). Metaphor and catachresis. In J. Bender & D. E. Wellbery (Eds.), *The ends of rhetoric: History, theory, practice* (pp. 60–73). Stanford, CA: Stanford University Press.

Petersen, A. (2001). Biofantasies: Genetics and medicine in the print news media. *Social Science & Medicine, 52*(8), 1255–1268.

Pollack, A. (2012, July 12). That fresh look, genetically buffed. *The New York Times.* Retrieved January 28, 2015, from http://www.nytimes.com/2012/07/13/business/growers-fret-over-a-new-apple-that-wont-turn-brown.html?pagewanted=all&_r=0.

Pozzer-Ardenghi, L., & Roth, W. (2004). Making sense of photographs. *Science Education, 89*(2), 219–241.

Reynolds, G. (2005, September). Will we grow babies outside their mothers' bodies? *Popular Science,* 72–78.

Rose, S. (1995). The rise of neurogenetic determinism. *Nature, 373*(6513), 380–382.

Rusting, R. L. (1992, December). Why do we age? *Scientific American,* 130–141.

Sachs, J. S. (2008, February). This germ could save your life. *Popular Science,* 64–69, 90, 92, 94.

Saey, T. H. (2010, July). All patterns great and small: Researchers uncover the origins of creatures' stripes and spots. *Science News,* 28–29.

Stover, D. (2002, January). Looks can be deceiving. *Popular Science,* 74–77.

Treise, D., & Weigold, M. F. (2002). Advancing science communication A survey of science communicators. *Science Communication, 23*(3), 310–322.

U.S. Food and Drug Administration. (2014). *Genetically engineered animals.* Retrieved April 23, 2015, from http://www.fda.gov/AnimalVeterinary/DevelopmentApprovalProcess/GeneticEngineering/GeneticallyEngineeredAnimals/default.htm.

Väliverronen, E. (2004). Stories of the "medicine cow": Representations of future promises in media discourse. *Public Understanding of Science, 13*(4), 363–377. doi:10.1177/0963662504046635.

van Dijck, J. (1998). *Imagenation: Popular images of genetics.* New York: New York University Press.

Wright, S. J. (1999, July–August). Human embryonic stem-cell research: Science and ethics. *American Scientist,* 352–361.

3

The Microscopic View: Minuscule Science and Art

When genetic research shifted its focus from observable physical traits to cellular and molecular phenomena, micrographs rose to be an important visual genre in public communication. Appearing in popular science magazines since around the 1950s, micrographs are routinely used to materialize scenes and objects invisible to the unaided eye. In one way, micrographs are a type of photograph, ones that are taken via microscopic instruments to provide magnified views. At the same time, the microcosmic world depicted by micrographs is distinctly different from the familiar, physical world captured by regular photographs. As such, micrographs present unique trends, promises, and complications in communicating genetics to public audiences and deserve a separate discussion in this chapter.

Micrographs and Micrographic Illustrations

While it is not the purpose of this chapter to be a historical study, it is worth noting, for context, that the history of modern microscopes dates back to at least the seventeenth century, when optical lenses were used to enlarge objects under observation. At the time, the technique of

© The Author(s) 2017
H. Yu, *Communicating Genetics*,
DOI 10.1057/978-1-137-58779-4_3

photographing a microscopic observation was not yet available. To publicize what they saw under a microscope, observers drew their observation and produced what is known as a microscopic illustration. Some of the most well-known such illustrations were created by English natural philosopher Robert Hooke (1635–1703) and preserved in his collection *Micrographia*, which was published by The Royal Society in 1665. *Micrographia* contains many graphically detailed and dramatic microscopic illustrations, of which the most well known is probably "The Flea" (Fig. 3.1). Printed as a foldout, "The Flea" is larger than the size of the book, measuring at about half a meter in length (Ford 1993, p. 174).

As the technique to directly photograph microscopic views became available in the nineteenth century (Overney and Overney 2011), actual micrographs began to appear in biological studies. Pursuing the same specimen as Hooke, English surgeon Arthur Durham used an optical microscope to capture the image of a flea in 1863 (Fig. 3.2 left); a decade later, German physician and microbiologist Robert Koch published several micrographs of the bacterium *Bacillus anthracis*, the agent that causes anthrax (Fig. 3.2 right).

Although Durham's flea came 200 years after Hooke's, the latter, in many ways, *appears* more real. In Hooke's work, one can more clearly

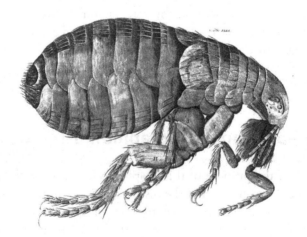

Fig. 3.1 Robert Hooke's 1665 microscopic illustration "The Flea" (Hooke 1665, scheme 34)

Fig. 3.2 *Left* Arthur Durham's 1863 micrograph of a flea (National Media Museum 2016); *Right* Robert Koch's 1877 micrograph of *Bacillus anthracis* (Koch 1877)

see the flea's mouthpieces and eyes, its scale-like skeleton structure, the demarcation of its body parts, and the hair and bristles on its body and legs. The shadings used in the illustration also convey a sense of body texture that the translucent micrograph fails to suggest. Given the time Hooke lived in, this superior magnification is not likely caused by him having access to the more sophisticated microscopic instrument. The perceived vividness and clarity of his flea, then, is largely a result of Hooke's (educated) speculation and creative invention. Reflecting what Galison (1998) called the pre-nineteenth-century paradigm of metaphysical images, Hooke's flea is an idealized depiction of nature, created by a "genius" scientist who has, presumably, seen many a flea and used his observation and creativity to record the "perfect" flea as nature intends it to be. Even to today's readers, Hook's flea remains a striking view.

But what might be celebrated in the seventeenth century as a perfect creation by a genius ceased to be so around the mid-to-late nineteenth and early twentieth century when the scientific visual paradigm shifted from metaphysical images to mechanical images (Galison 1998). Under the new paradigm, which emphasized mechanical objectivity, idealization "became anathema," and drawings, as a whole, were deemed "erroneous" (Galison 1998, p. 332). "Untampered" micrographs, as photographs (see Chap. 2), were deemed superior visual evidence because they were supposedly the product of automatic, mechanical processes free of human interference.

Given this context, it is curious to see that early twentieth-century popular communications of genetics still favored microscopic illustrations. Figure 3.3 shows one example, depicting the cells of an onion root undergoing mitosis, a process wherein a parent cell divides into two genetically identical daughter cells. At a first glance, Fig. 3.3 resembles a micrograph: Much as Fig. 3.2, it has a grainy background, fuzzy details, and organic shapes. But upon closer examination, one notices that the cells, including their membranes and nuclei, are well defined and neatly arranged—too defined and arranged for a natural specimen. The image, then, appears most certainly a microscopic illustration. Although it may not look as visually striking as Robert Hooke's flea, it is similarly meticulous, precise, and idealized—a perfect image that seems more real than the irregular and ambiguous reality.

What is even more curious is that in his description of the image, the author seemed, consciously or otherwise, trying to present the image *as* a real micrograph. The original caption reads, without additional context, that the image is "highly magnified," which, in the presence of the image, readily suggests a direct, magnified view. Furthermore, in

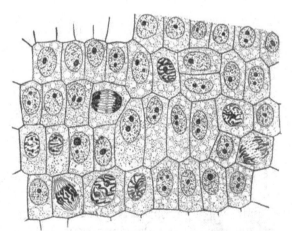

FIG. 2.—VERY THIN SECTION FROM THE TIP OF A GROWING ONION ROOT, SHOWING HOW THE CELLS OF MULTICELLULAR ORGANISM ARE BOUND TOGETHER.
Highly magnified. Several of these cells are in process of self-division.
(From Wilson's "The Cell.")

Fig. 3.3 Microscopic illustration of the tip of a growing onion root (Lane 1908, p. 45)

the original article, the author repeatedly emphasized that the image *is* what one actually sees and *is* the reality: "[It is] an excellent picture of *the sight one sees*"; "Here we *see* the cells bound together, each with its nucleus and germ-speck"; "Here and there in the drawing is *seen* a cell in which the nucleus is replaced by odd looking black, curved rods, or loops" (Lane 1908, p. 45, emphasis mine).

Between a pristine micrographic illustration and an evasive emphasis on reality, there seems an uneasy tension in this early twentieth-century account of mitosis. The author seemed consciously aware that a micrograph *would* have been a preferred visualv choice, if only a "good" micrograph existed. But therein lies the problem: Conclusive observation and evidence of cellular level genetic activities were simply not available at the time, due in part to the lack of high-power microscopes. As Wilson (1913) wrote, rather tentatively, of the state of understanding at the time:

> *Reference has already been made* to the fact that *at a certain period*, shortly before the germ cells are formed, corresponding maternal and paternal chromosomes become coupled in pairs, side by side (synapsis). This process is always followed by *a more or less* intimate union of the two threads, *perhaps in some cases* by actual fusion. *The evidence is still more or less conflicting as to exactly what follows*, but it is certain that at a later period two separate and parallel threads again become distinct, and these *may* separate so as to pass unchanged into different germ cells. These two threads are *believed by many observers* to be identical with those that originally united in synapsis, but *this is in dispute*. (p. 80, emphasis mine)

If trained researchers were uncertain of what they observed under the microscope or how to interpret those observations, any such micrographs would not function as accessible evidence for public readers. In fact, they could invite alternative readings that undermine the ethos of an author. A microscopic illustration, on the other hand, removes ambiguities, inconsistencies, or what an author believes unimportant details while highlighting what is believed to be the most accurate and salient interpretations. These considerations could explain popular communications' persistent use of microscopic illustrations in the earlier decades of molecular genetic research, especially illustrations that, as Fig. 3.3, look like micrographs (e.g., Thone 1927; Ruckes and Mok 1931; Kiesselbach 1951).

Viewed this way, these illustrations are not only scientific evidence in the narrow sense but interesting rhetorical artifacts. They are necessary, savvy even, rhetorical moves on the part of earlier geneticists to publicize emerging but inconclusive findings while sustaining and promoting their credibility. By this, I do not mean "real" micrographs are somehow non-rhetorical or not bound up with the ethos of science. It is common practice, for example, for a researcher to take multiple micrographs from repeated experiments in order to construct the "strongest" evidence. But as with the case of photographic versus illustrative electrophoresis images (see Chap. 2), the illustration genre in both cases affords more "artistic leeway."

On the other hand, it is also possible to question whether these microscopic illustrations are semiotically excessive. If the understanding of mitosis at the time was so uncertain, is it acceptable to represent that uncertainty in such definitive visual forms? Should readers be presented instead with fuzzy micrographs so they could witness the uncertain nature of the event and make their own interpretations? After all, vision is such a privileged sense in the Western culture that "seeing" is often equated to "knowing" (as in "Seeing is believing" or "I know it when I see it"). If the material reality of the "seeing" does not exist, then the portrayed "knowing" is excessive at best and misleading at worst.

These different interpretations of microscopic illustrations, in a way, reflect two alternative approaches to the public communication of science. The first approach, what may be termed the "need to know" approach, concedes that an enormous amount of information is involved in any scientific topic and further holds that public communications of science should focus on what the publics need to know, not everything that would be nice to know (Fischhoff 2013). Overwhelming the public with information, it is believed, undercuts their trust of science, hinders their understanding, and complicates decision making (Fischhoff 2013). According to this view, visual evidence is best filtered based on expert judgment so that the "need to know" elements are highlighted while the others are removed for overall easier access.

Despite its good intention and rationalized reasoning, the "need to know" approach hands over the gatekeeper role of science communication to scientists and their associated experts. Although the "need" analysis is to be formally based on public concerns, it is also fundamentally driven by the goals and desires of the gatekeeper (see Fischhoff 2013). When the topic under discussion is one of controversy (as in the case of GMO) and when public concerns and gatekeeper goals conflict, the "need to know" approach allows institutions to unilaterally determine what is or is not relevant to communicate to the publics. The non-required GMO product labeling in the USA is a case in point.

Contrasting with this approach is what may be called the "right to know" approach, one that emphasizes, as a principle, the publics' right to define the problem, access information, and generate knowledge (Lynn 1990). In the USA, this approach had its root in citizen participation in workplace and community risk management (Lynn 1990). The same principle is used by public initiatives to lobby state legislations to require GMO labeling. Recently, such efforts resulted in the passing of the GMO food labeling bill in Vermont, the first state in the USA to pass such a law. Based on the "right to know" approach, visual evidence is best presented "as is," including its full range of ambiguities and imperfections, and non-expert readers are trusted to be able to deal with such uncertainties and to convert the information into practical and actionable knowledge.

Despite *its* good intention and social justification, the "right to know" approach idealizes the power of information possession and romanticizes publics' interests in and abilities to process scientific information. In communicating such abstract and specialized topics as genetics, an adamant adherence to this approach can result in communication products that are so full of background information and knowledge that they seem no different than communications that have not at all considered public readers' needs or rights.

There is, then, a third approach, not necessarily compromised but eclectic. An instance of such an approach can be seen in Fig. 3.4. The left panel of the figure shows a micrograph of a bacterium fossil found from 3.5-billion-year-old rocks, revealing the earliest known form of life that existed on earth. The middle panel then illustrates the microfossil in further detail, presenting the bacterium in somewhat regular segments. These segments are not obvious in the microfossil but appear reasonable once illustrated. Finally, the right panel shows a micrograph of a cyanobacterium, which represents a group of present-day, highly evolved bacteria; in this case, even without the help of illustrative tracing, the micrograph conveys clear segmentation. Using this series of visual evidence, the author argued that the two life forms, 3.5 billion years apart, are remarkably similar.

Figure 3.4 thus leverages microscopic illustrations' ability to highlight selective information as well as micrographs' ability to make available a fuller range of information. By showing as well as illustrating the microfossil in question, the author provided readers with information that would be "nice to know" for contemplation while demonstrating his version of what needs to be known. This way, visual ambiguity is not sidestepped, while the cogent argument is maintained. In particular, the process of how visual evidence is filtered by scientists is revealed, and

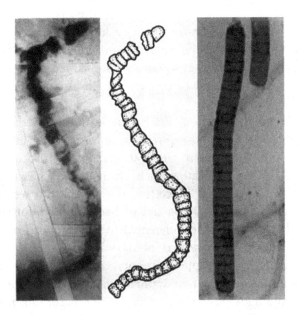

Fig. 3.4 Combined use of micrographs and microscopic illustration (de Duve 1995, p. 430). Courtesy of Dr. J. William Schopf

readers are afforded the chance to assess that interpretive filtering. When microscopic evidence is less than apparent, such seems a more genuine way to invite public scrutiny, conversation, and trust.

The (Elusive) Microscopic Evidence: Syntactics, Semantics, and Pragmatics

Microscopic illustrations, as seen in Fig. 3.4, continue to find use in contemporary popular communication of genetics. But as genetics entered the DNA age and especially as microscopic technologies advanced with the refinement of optical microscopes and the rise of high-power electron microscopes, "real" micrographs, by all measures, became the dominant genre for minuscule evidence post-1950s.

Earlier such micrographs were usually unadorned, black-and-white images. And in those earlier decades of the DNA age, chromosomes were a common subject. These images frequently compare normal and aberrant compositions of the human chromosomes, correlating the latter to genetic diseases. A number of them (e.g., Epstein and Golbus 1977; Patterson 1987) zero in on Down syndrome. As Fig. 3.5 shows, humans have (or should have) 22 pairs of autosome chromosomes (labeled No. 1 through No. 22) and one pair of sex chromosomes (labeled X and Y). Two X chromosomes, as in the case of Fig. 3.5, identify a subject as female; one X chromosome and one Y chromosome identify a subject as male. The chromosomes were stained to bring out their characteristic bands, which correspond to gene clusters. What is significant in this case is that instead of having two copies of No. 21 chromosome, the individual has three copies, what is known as a trisomy 21. This aberration is what causes Down syndrome.

Figure 3.5 and similar micrographs, in a way, took after where photographs left off when genetics shifted its focus from physical observation to molecular-level activities. Just as photographs were a convenient and convincing genre to document macro-level inheritance and mutation in the age of classical genetics (see Chap. 2), micrographs are trusted to record life's microcosm in the DNA age. Micrographs' ability to magnify and make visible what is otherwise invisible gives them

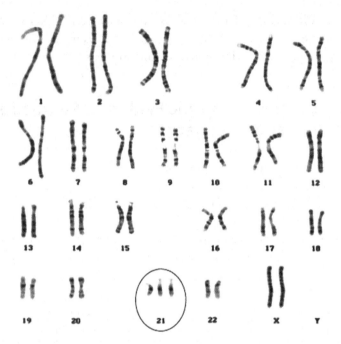

Fig. 3.5 One extra copy of No. 21 chromosome causes Down syndrome. Adapted by permission from Macmillan Publishers Ltd: Nature Reviews Genetics (Antonarakis et al. 2004, P. 734), copyright (2004)

obvious value in conveying minuscule evidence. Moreover, in the public communication of genetics, their significance goes beyond providing apparent evidence. By making what is invisible (and thus incomprehensible) visible (and thus potentially comprehensible), they help to demystify genetics, its object of study, and its approach. In Price's (1996) words, the act of seeing becomes part and parcel of knowing. Though any such published micrographs are, no doubt, results of careful design and selection by a scientist, they are, for all intents and purposes, "firsthand" visual evidence that affords public readers the opportunity to examine and evaluate research findings.

Of course, the physical pixels on a micrograph do not automatically "make sense." While this is true of all visual representations, it is especially the case with micrographs, as a semiotic analysis of the genre would suggest. According to semiotics, signs can be examined from three aspects that, together, map how humans make sense of them. First, the syntactic aspect is interested in how signs are formally constructed (with lines, shapes, etc.) and how viewers perceive that construction; next, the semantic aspect considers how signs signify objects and how readers recognize those objects; last, the pragmatic aspect examines how signs are interpreted in their wider social–cultural contexts and carry a certain message for viewers (Goldsmith 1984; Dragga 1992).

Syntactic Consideration

At the syntactic level, micrographs often seem to consist of unorganized grainy particles as opposed to defined lines or regular shapes that one generally encounters in the physical world. For readers without prior experience with this visual genre, then, perceiving *that* certain objects are purposefully depicted requires conscious effort. Take as an example Fig. 3.6, which is an electron micrograph that depicts an exploded *Escherichia coli* cell and its scattered genetic materials. The micrograph features a distinct dark area in the upper left corner, which the caption

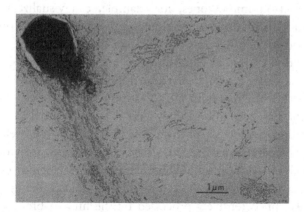

Fig. 3.6 Micrograph of an exploded *E. coli* cell. Because of copyright restrictions, the original micrograph published in Miller's (1973) *Scientific American* article cannot be reproduced here. The image reproduced here captures the same content and is from another publication by the same author and colleagues. From Miller et al. (1970, p. 393). Reprinted with permission from AAAS

explains to be the remains of the *E. coli* cell wall after the cell was exploded by osmotic shock. With this explanation, one can reasonably make out a unified image in that dark area.

The caption proceeds to explain that when exploded, the cell "extruded cellular contents consisting of fine fibers with fibrillar segments and attached strings of granules. The fibers are portions of a bacterial chromosome, the strings of granules are polyribosomes, and the fibrillar segments are ribosomal genes" (Miller 1973, p. 41). At this point, the minuscule evidence becomes difficult to follow: The micrograph supposedly features "fine fibers," which are thin, elongated objects; "strings," which are also thin, elongated objects; and "fibrillar segments," which are segments of fibers, so once again, thin, elongated objects. Given these visual overlaps and for readers who do not have prior experience observing such elements as chromosomes and polyribosomes under the microscope, the intended "look *at* this and that" activity easily becomes a frustrating "look *for* this and that" exercise.

Some readers may suspect that, Fig. 3.6 being a 1970s production, its syntactic issues are due to its modest magnification and plain presentation. Indeed, as microscopic technologies advanced, so did the resolution of micrographs and the sophistication of their appearance. Today's scanning electron microscopes, for example, can visualize organelles' surface structure in three-dimensional appearance, and colors can be easily applied to help differentiate intricate details. That said, advanced technical and visualization capabilities do not in themselves erase accessibility issues. Color is a particular case in point.

While earlier micrographs mostly appeared in grayscale, since the 1980s, color has been increasingly and is now routinely used. From an information design perspective, this change is to be welcomed as colors are a superior viewing aid. They can highlight areas of interest in a micrograph without employing extra markers (such as callouts or arrows) that may interfere with the minute objects being observed. They are also pre-attentively processed by the human brain at a lightening speed, thanks to the sensitive cone cells in our retina, and enhance pattern recognition (Ware 2012). The colors red and green, in particular, create a heightened visual contrast,[1] and frequently, dyes and fluorophores used in microscopic examinations result in these two colors. For example, Fig. 3.7, reported in a Science News article (Saey 2010), employs them to demonstrate the wing pattern of a fruit fly species known as *Drosophila guttifera*. The pattern, induced by a morphogen and associated pigmentation gene expression, consistently includes 16 black spots and four gray shadows, which are made to fluoresce green[2] and red, respectively. The two colors are bright, unmistakable, and highly distinct.

All these advantages of color, however, cease to apply when we consider the needs of viewers who have various forms and degrees of color deficiencies (commonly known under the imprecise, blanket term "color blindness"). The two most common forms of deficiencies are protanopia and deuteranopia, both of which are, precisely, red–green deficiencies; the former is caused by a lack of red cone cells on the retina and the latter by the lack of green cone cells. For protanopia and deuteranopia viewers, the vivid micrograph in Fig. 3.7 become those shown in Fig. 3.8, where all colors turn into some brownish yellow and offer minimal visual contrast for pattern recognition.

Figure 3.7 is far from being an exception, as similar red-green-contrast micrographs were frequently encountered in this study (see, e.g., Nettelbeck and Curiel 2003; Saey 2009; Gage and Muotri 2012). It is also useful to note that it is not just pure red and green that can cause problems but potentially any color that contains red and green pigments (for example, orange, which mixes red and yellow). Moreover,

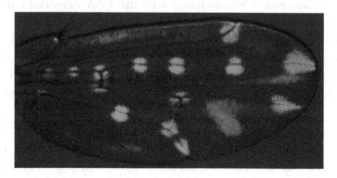

Fig. 3.7 Micrograph uses contrastive red and green to illustrate fruit fly wing pattern. Adapted by permission from Macmillan Publishers Ltd: Nature (Werner et al. 2010, p. 1144), copyright 2010 (color figure online)

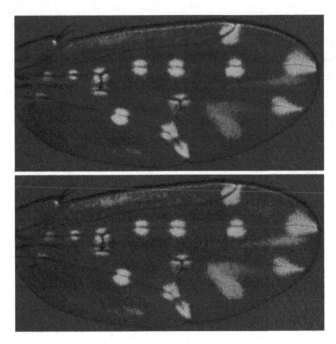

Fig. 3.8   *Top* Protanopia view of Fig. 7; *Bottom* Deuteranopia view of Fig. 7. Adapted by permission from Macmillan Publishers Ltd: Nature (Werner et al. 2010, p. 1144), copyright 2010 (color figure online)

a substantial number of people are affected by color deficiencies: About 10% of males and 0.5% of females in the USA have color deficiencies; among people of Northern European descents, 8% of males and 0.5% of females have red–green deficiencies (Color Universal Design Organization 2006; Wong 2011). Given these, color is not, as some may think, a minor syntactic consideration. Its use is part of the complex semiotic framework and design choices that impact the perceived effect of micrographs.

This, of course, does not mean that we should avoid color and go back to the pre-1980s grayscale presentation. It does mean that in cases where grayscale does the job adequately, as in Fig. 3.5, there is no need to use color just because one can. And when colors *are* employed to enhance presentation, visual imaging programs should be used to change problematic color channels used in an original experiment into ones that are less troublesome for color-deficient viewers (Wong 2011; Okabe and Ito 2008). For example, blue is often considered a "safe" color because "blue deficiency" is very rare. Or, in red–green contrast, red can be replaced with magenta, which is a mix of red and blue so viewers who have difficulty discerning red pigments can recognize the blue.

Semantic and Pragmatic Considerations

As mentioned above, after viewers perceive that certain elements are presented in a micrograph via syntactic clues, semantic recognition and pragmatic sense-making must follow. Here again, micrographs present unique challenges.

Consider, for a moment, the process of viewing a photograph from everyday life, say one of cherry blossoms. Such a viewing process is often taken for granted because it is frequently rehearsed, but it is not a simple or automatic process. From the semantic perspective, a viewer must rely on prior experience with cherry trees or other comparable objects to recognize what is photographed—if one cannot, say, tell the flowers from the leaves from the branches, one is not "seeing" any cherry blossoms. After recognizing the image, a viewer then must possess the relevant social and cultural frameworks if she is to interpret the image as a sign of nature or beauty. For viewers who do not have corresponding semantic and pragmatic frames of reference (imagine viewers who come from terrains without blooming trees or viewers who rely on cherry trees as a source of income), recognition would not be instantaneous and interpretations would be different.

The same process applies to viewing micrographs. When a scientist audience examines micrographs, she relies on, among other things, knowledge of cell structures (and perhaps an appreciation of such structures), experiences with cell staining and sample preparation (and perhaps an appreciation of the careful execution of such processes), experiences of how different types of microscopes render observations, and awareness of terminology and classification systems (Dinolfo et al. 2007).

With these knowledge and experiences, and then given a micrograph such as Fig. 3.9 on Streptococcus bacteria, which was published in *Popular Science*, a scientist audience can make reasonable recognition and interpretation with little additional prompt: These bacteria are responsible for various common infections; the bacterium in the middle of the image is dividing to reproduce; the green areas inside the bacteria represent higher-density materials, likely the bacteria's DNA; and (possibly) the micrograph was well prepared and should be valuable in the study of the bacteria. If given more details about the image being a resistant pyogenes strain, this audience may further consider the deadly necrotic fasciitis caused by this fast-spreading strain and recognize the micrograph's added intrigue and value for scientific research.

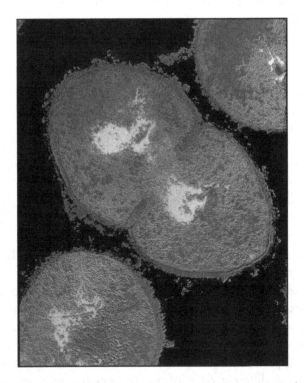

Fig. 3.9   Micrograph of the Streptococcus bacteria without context ("A plague," 1996, p. 53). Credit: Dr Kari Lounatmaa/Science Photo Library (color figure online)

However, for public readers who do not have existing semantic and pragmatic frames of reference, Fig. 3.9 does not conjure up such visual recognition and interpretation—or, possibly, any recognition and interpretation. What one sees is a close-up view of some colorful organisms, so close-up that the visual context is lost. One may well be told, through a caption or body text discussion, what these organisms are. But the purpose of looking at these organisms—the kind of purpose that we can more clearly discern in Figs. 3.4, 3.5 and such—is absent in the image proper.

Micrographs like Fig. 3.9, I argue, constitute unsubstantial visual evidence; that is, their function does not go much beyond the basic act of showing "this is X as discussed herein." These contextless micrographs are, again, not uncommon in this study (e.g., Gibbs 2003; Bakalar 2006; Beil 2012). Granted, these colorful and novel images can add visual and emotional interest to the page, a point that I discuss in detail in the next section. But for micrographs to serve as meaningful and worthwhile visual evidence in popular communication, they also need to be cognitively interesting and afford readers with necessary semantic and pragmatic scaffolds. By this, I do not mean that they should instill in public readers the kind of formal knowledge and experience scientists possess; rather, I propose that they leverage readers' existing knowledge and experience to construct familiar, relevant, and accessible reference frames.

One such reference frame is storytelling, or narratives. As scholars in anthropology, psychology, literature, and education demonstrate, narratives are a primary thought and discourse process for people to make sense of experience, to build knowledge, and to inform, persuade, teach, and instruct (Connelly and Clandinin 1990; Perkins and Blyler 1999; Jameson 2000; Altman 2008). Though these scholars are primarily speaking of verbal stories as rooted in mythologies and folklores, the same, I argue, is true of visual narratives. Images have been used to tell stories for millennia: painted on cave walls, hung on medieval tapestries, printed on paper, and very recently published on screen (Cohn et al. 2012).

What, then, constitutes a story or, for our purpose, a visual story? According to Edward (1997), a story should employ four basic elements: character, setting, narrator, and plot.[3] Based on this framework, Fig. 3.9 easily possesses three narrative elements: The Streptococcus bacteria assume the role of characters whose state or action is the interest of observation; the cellular environment within the human body where the bacteria exist provides the setting; as for the narrator, it is understood to be the scientist who undertook the observation and procured the image or the communicator who presented the image. What is missing and preventing the figure, and similar context-less micrographs, from creating a coherent visual narrative is the fourth element: plot.

Plot, as Brooks (1992) wrote, is "an embracing concept for the design and intention of narrative, a structure for those meanings that are developed through temporal succession, or perhaps better: a structuring operation elicited by, and made necessary by, those meanings that develop through succession and time" (p. 12). Put simply, it is a certain logic developed through successive events, actions, or states of being. With regard to visual narratives, these successive events, actions, or states would be rendered by discrete visual components, whose correlation to each other then creates different logics and perceived plots. Given different visual successions, a range of logics are possible, which cannot be exhausted in the limited space here. What follows, then, are common plots used by micrographs encountered in this study.

The first such plot is "danger," which is not surprising given the contemporary media portrayal of genetics as a medical human drama (see Chap. 2). For example, when micrographs depict virus particles within a host cell (Stemmer and Holland 2003) or AIDS virus surrounding a T cell (Hoffman 1994), the succession of visual elements (virus particles and then the host cell; AIDS virus and then the T cell) cue readers of impending danger—much the same way movies do by presenting shots of unsuspecting victims followed by those of plotting villains. Operating along the same mechanism but in the opposite direction is the plot of "rescue." This plot is enacted when micrographs depict white blood cells approaching and attacking cancer cells (Cohen et al. 1999) or when they depict telomeres "bracketing" chromosomes to prevent the latter from tangling up (Travis 1995). It should be noted that with the contextless micrographs mentioned earlier, similar plots of danger and rescue may often be intended. In the case of Fig. 3.9, presumably, the underlying story is that Streptococcus bacteria pose danger to human health. However, from a visual standpoint, such a story is not made apparent because of the images' lack of context, which reduces them to perfunctory visual labels as opposed to engaging evidence.

While danger/rescue stories are often told in single-frame micrographs that embed successive visual elements within one same panel, another common plot is frequently told in multi-frame, serial micrographs: the plot of "change," a plot common in both literary and real-life stories. The change plot can be further divided into "changing for the better," or a story of progress, and "changing for the worse," a story of deterioration. The former is illustrated, for example, by a series of micrographs that captures the growth of neuronal axons in zebra fish embryos (Dahm 2006). In this case, progress is visually narrated through the gradual growth that happened between each successive view, reinforced by the time stamps on the micrographs.

On the converse side, Fig. 3.10 tells a story of deterioration. The left frame shows the germ cells from the testes of juvenile male alligators that were born in a clean lake; the right frame shows the same cells from juvenile male alligators that were born in a lake with high

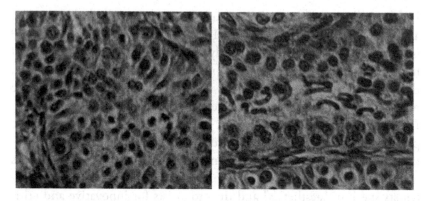

Fig. 3.10 Multi-frame micrograph tells a story of deterioration caused by environmental hormones (McLachlan & Arnold 1996). Reproduced with permission from *Environmental Health Perspectives* (Guillette, Jr. et al. 1994, p. 686) (color figure online)

concentrations of DDT and other pesticides. Between the two frames, one witnesses changed habitats giving rise to changed cellular structures: alligators from the polluted lake develop abnormal, bar-shaped cell structures that are not found in alligators from clean environments. These structures, the article argued, are evidence of environmental hormones' adverse effect on animals' gene expression and resultant genital defects.

In advocating for visual storytelling, I do not necessarily suggest that all micrographs be framed as a dynamic story. A "static" presentation may be suitable if the purpose of a micrograph is to demonstrate the appearance or structure of an organelle. But with such a purpose, visual markings (such as colors or callouts) would be needed for readers who do not have relevant frames of reference to recognize and interpret the organelle in question. Indeed, when such markings are used and thereby segment a single micrograph into successive visual elements (as in the case of fruit fly wing spots and shadows in Fig. 3.7), it is possible to argue that the succession gives rise to a plot of "uniting"/"dividing" and again makes the micrograph a coherent story.

Works of Art, or Visual Advertising

Holding what may be called a traditional approach to science visualization, Gross and Harmon (2013) asserted that "whether a visual is good, bad, or ugly from a purely artistic perspective is a matter of no particular consequence from a purely scientific perspective. Indeed, the vast majority of contemporary scientific visuals are as plain, some might say relentlessly drab, as the text that accompanies them" (p. 10). I am not so sure that there can be a "purely" artistic or "purely" scientific perspective when it comes to contemporary popular science visualization or, indeed, science visualization in general. At a time when artists working in various media seek inspiration from science and when scientists speak of "aesthetics" and turn to artists for innovative and critical new perspectives, what is artistic and what is scientific have started to blend (see Ede 2002; McGhee 2010; Kjærgaard 2011). "Micrograph art," as a case in point, is now a common term.

This blending also makes sense from the alternative perspective of economics. Scientific institutions and individual scientists seek promotion in the public domain for research support and funding; media reports of science, as any other communication products, depend on subscription and sales for continuation. For either party, attracting viewer/consumer attention is of primary importance. Micrographs that are used to show unfamiliar—and therefore novel—microorganisms without context, as mentioned earlier, already belie an attempt to create visual appeal. This attempt becomes magnified (pun intended) when we consider some of the contemporary micrographs accompanying popular communication of genetics, micrographs that share distinct strategies with advertising, marketing, and entertainment industries.

In examining these micrographs as objects of art or advertising, I am not merely interested in their ability to catch readers' attention, though that is certainly part of it. Rather, I am interested in the overall, implicit semiotic meanings these micrographs create. As Barthes (1957) wrote, a successful semiotic object functions as a myth: the image may seem to *naturally* conjure up a certain concept or give *foundation* to the concept without the viewers consciously realizing it (p. 129). As shown below, in our case, micrographs may lead public readers to *naturally* conjure up certain mentalities and value judgments about genetics and about science in general. Teasing apart these supposedly natural affects allows us to fully understand the motivation and social impact of popular science micrographs.

Mysterious Splendor: An Aesthetic of Science Visualization

Given the research focus of modern genetics, relevant micrographs often depict microorganisms and molecular organelles, which assume intricate structures very different from what one sees in the physical environment. As such, and with the aid of color and high definition, these micrographs can easily create a sense of mysterious splendor. Figure 3.11 is an example, which was prepared by the National Institute of Allergy and Infectious Diseases after the Ebola outbreak in 2014.

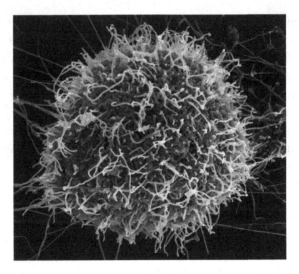

Fig. 3.11 Scanning electron micrograph of Ebola virus particles budding from an infected cell (Collins 2014) (color figure online)

It accompanies a National Institutes of Health report on scientists' attempt to sequence the Ebola genome in order to identify its origin and evolution and aid the development of diagnostics, therapies, and vaccines.

In Fig. 3.11, Ebola virus particles, in their curious filament shape and a bright color of sky blue, are seen budding out of an infected cell. Some of the virus filaments are curly and crawl on the surface of the cell; others are stretchy and project out from the cell surface, creating a tangled web. The cell, yellowish in color, is itself a curious object with an unusual, wiry texture. The image as a whole is chaotic with the numerous virus filaments yet also strangely orderly with the virus enclosing the infected cell. It is, in short, mysteriously splendid. This visual impression, in Barthes' (1957) words, *naturally* gives a similar *foundation* to genetic research as that which explores and conquers the mysteriously splendid. Of course, verbal metaphors of biotechnology as an adventure and expedition also exist (see van Dijck 1998). But visual representations serve to justify the speech in excess (Barthes 1957) and offer a more concrete, hence more "real," imagery of genetic research as exciting and superior—and worthy of publics' uttermost interest and support.

This mysterious splendor approach—compared with other approaches discussed below—is the most common way visual creators choose to present micrographs in contemporary popular reports of genetics. In *Popular Science*, for example, the recurring megapixels section frequently features stunning micrographs, from a beautiful look of parasitic flatworms with their stem cells to fired-up neurons after a genetic tweak. Similar micrographs are also the mainstay when one thinks of "micrograph art" or more generally of the intersection between science and art. As a notable example, the National Science Foundation (NSF) co-sponsored visualization challenge Vizzies frequently awards winning titles to such micrographs: a mesmerizing set of color-coded concentric rings that are the cells of a mouse's eye or translucent, shimmering droplets that are the fine hairs of a young cucumber (see National Science Foundation 2015a).

To a large extent, such micrographs are popular because mysterious splendor can be easily and unambiguously interpreted as beautiful and intriguing. And being beautiful and intriguing is a convenient, though under-articulated, view of the aesthetic of science visualization[4]. This view holds such factors as color, novelty, and visual "nicety" in high regard but does not necessarily engage with questions of overall semiotic intent and effect (see Kjærgaard 2011; Stone 2009; National Science Foundation 2015b). What the aesthetic ends up achieving, then, is promoting science and cultivating public appreciation rather than complicating the nature of science and encouraging deeper public engagement.

Fear and Sex: Ways of Modern Advertising

A widely held belief in visual design is that humans are hardwired into paying attention to certain images: notably, images of food, sex, danger, and people's faces (Weinschenk 2011). The advertising industry has capitalized on these items in order to create maximum viewer attraction, and parallel moves, surprising enough (or maybe not so much after all), are seen in the contemporary micrographs encountered in this study.

The first notable parallel is the evocation of fear. Fear is a stimulation that humans respond strongly to, thanks to the evolutionary demand that links survival to intense and fast response to danger (Carretié et al. 2005). Even though researchers disagree on how much fear is "optimal" and acknowledge the influence of other factors, generally, studies show that fear activation increases viewer recall of advertisements and positively influences their acceptance of and attitude toward advertisements (LaTour et al. 1996; Snipes et al. 1999; Kim et al. 2014). Because of this, fear appeal is a common strategy in marketing and advertising. By first presenting fear (e.g., fear of vehicle accidents) and then offering a product or action that promises to dissipate that fear (e.g., cars with certain safety features), this strategy persuades consumers to at least pay attention and at best purchase the product or adopt the action.

Contemporary media reports of genetics use micrographs in similar manners. With their ability to magnify minuscule objects, especially objects of an unfamiliar or questionable nature such as cancer cells, bacteria, and insects, micrographs can be easily used to evoke fear (see, e.g., Travis 1997; Cornelis et al. 1998; Mone 2003; Saches 2008). In stating this, my point is not simply that these objects are fearsome in nature—after all, what objects become the focus of scientific observation is, for the most part, a given and a question of research needs rather than a communication strategy. My point is that one same object can be presented in multiple ways, and some micrographs encountered in this study make conscious semiotic choices that amplify fear.

Take Fig. 3.12 as an example. It shows an *Anopheles* mosquito, which can spread malaria. The image shows, emphatically, the less-than-pleasant (to say the very least) details of the mosquito, with foreground focus on its mouthpieces, compound eyes, bristles, spidery legs, and rugged body. The mosquito is further portrayed in a menacing posture with legs and wings spread out and head and mouthpieces facing the reader, as if ready to attack. The micrograph features a drab, brown color and a gritty texture that adds to the general unpleasantness. These semiotic choices convey feelings of disgust and fear independent of the fact that the creature depicted can spread the deadly disease of malaria. In similar fashions, objects such as cancer cells are presented with gritty textures, sprawling filopodia (those slender cytoplasmic spikes that protrude from a cell), and drab colors to evoke visceral reactions (see, e.g., Cornelis

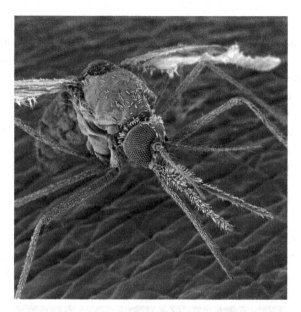

Fig. 3.12 A menacing *Anopheles* mosquito (Mone 2003, p. 41). Credit: Eye of Science/Science Source (color figure online)

et al. 1998; Karidis 2015).[5] And once again, such reactions are independent of the semantic meaning of the objects.

Precisely because they are unpleasant and fearsome, these images invite (or compel) one to pay attention, which helps to promote the media product in question. Moreover, the visceral reaction aroused may also be said to "sell" the genetic research being reported. That is, after the presentation of fear, genetic research is offered as a promising solution to dissipate that fear. In the case of Fig. 3.12, the accompanying report explained that a gene may be engineered into an *Anopheles* mosquito to knock out one of its essential genes. Thus engineered, the mosquito is released into the wild to mate with normal mosquitoes and pass its genetic defects to their offspring, starting a chain reaction of dying-off (Mone 2003). Genetic engineering is thus visually proffered as the ultimate "product" that conquers the mosquito, malaria, and fear. It is useful to note here that fearsome micrographs are hardly restricted to popular communications of genetics. Many supermagnified and graphic representations of cancer cells, viruses, and creatures like fleas and bedbugs adorn the Internet, including social media sites. Their popularity, I argue, derives from the same fear appeal. The fearsome micrographs encountered in this study, then, represent a larger and more prevalent media trend.

In addition to the fear appeal, another interesting (if less common, for lack of opportunity) parallel between modern advertising and science communication is sex appeal. Media industries have long recognized that sex sells. In Reichert and Lambiase's (2006) words, millennia of survival and pleasure hardwired certain shapes and images (buttocks and breasts, for example) into our brain as symbols of sex, and mass media use these images and connotations "to attract attention, to spark controversy in order to attract attention, to offer pleasure, and ultimately, to sell media products and consumer goods" (xiii).

Consider Fig. 3.13, which is a micrograph of the human sex chromosomes. Apparently popular, this image has been used in various media

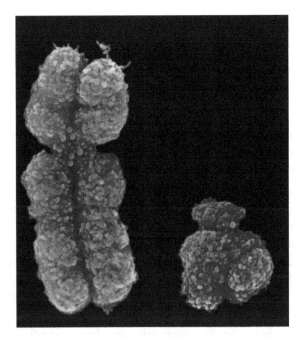

Fig. 3.13 Sensational sex chromosomes. Credit: Science Photo Library (color figure online)

reports, including a *BBC* news article on the possibility of a male mouse (or male human) remaining fertile despite of a damaged sex chromosome (Gallagher 2013); a *Scientific American* article that questions why the two sex chromosomes evolved into different shapes and structures (Jegalian and Lahn 2001); and a *Popular Science* (Trolio 2006) Q&A piece that ponders whether the male chromosome is undergoing continuous degradation and facing ultimate demise.

The apparent popularity of the micrograph, I argue, is due in no small part to the image's assumed sex appeal. As is typical of contemporary practice, the micrograph appears in full color: The male Y chromosome is blue, and the female X chromosome is pink. Given the many color-staining possibilities, these choices that correspond with the Western cultural perception of gender are not accidental. In addition, the two chromosomes are captured in such angles (of all possible angles) that their shapes, together with their suggestive colors, are ripe with a sexual connotation. The female chromosome is analogous in color and shape to red lips or, more strikingly, analogous in color, shape, *and* orientation to the vulva, the external female reproductive organ. The latter reading is reinforced by the frontal, voyeuristic view. When the female chromosome is thus suggested, it is inevitable that the male chromosome should conjure up parts of the male reproductive organ, namely, the two testicles, slightly different in size and one hanging lower than the other.

If we believe sex sells, then the image helps to promote not only the media products in question but possibly genetics as a research area intimately related to one of our fundamental needs. And once again, this sex strategy—at least with regard to the much more sexualized female bodies—reflects a larger media trend. SC Johnson's controversial public service advertisement on parental control, for example, uses a half-opened eye with lush eyelashes turned vertically to create the allusion of a vulva. Similarly, literary magazine *Granta*'s 2010 issue headlined "Sex" features an opened, double-layered pink purse turned vertically for the same visual effect.

Depending on who is asked and what perspective one holds, the micrographs discussed in this section may be interpreted differently: as an effective way to compete for viewers' attention and make science appealing and relatable, or, alternatively, as a questionable, profit-driven rhetoric used by scientists and science communicators to create publicity.

As Macdonald (1996) acknowledged (if somewhat reluctantly) in the context of science museums, public viewers are not captivated readers of science but "leisure-seekers" whose attention and interest are being vied for by multiple leisure avenues. Science communicators can no longer (if they ever were able to) assume that the content they create will be deemed edifying and eagerly consumed by publics; rather, as consumers, publics determine whether to tune into reports of science and if so, what kind (Macdonald 1996). The same applies to print-based science communication two decades later, with a vengeance. Given the competition from other media (from TV to video games to social media), let alone other leisure activities, popular science communication cannot and should not expect everyday readers who are not already fans of science to peruse its work, if that work does not take into consideration the readers' needs, wants, and interests.

With that in mind, the micrograph genre seems well positioned to spark publics' wonder, attention, and emotional interest, especially by bridging the age-old gap between arts/humanities and sciences. San Francisco-based artist Catherine Wagner, for instance, used scanning electron microscopy, together with other technologies, to create her exhibition "Cross Sections," which includes black-and-white imageries of various organisms. She appropriated, in one work, a cell division micrograph captured by scientists, "cleaned up extraneous 'noise' around the image and put it on a black background" to create a mesmerizing piece titled "Dividing Cell" (Cheng 2003). If we celebrate the visual appeal of such creations that are self-identified as art, it seems unreasonable not to celebrate similar creations in science communication just because they do not appear in an art gallery. And genetics, given its widespread social and cultural influence, offers an especially fertile ground to do cross-disciplinary work. As Anker and Nelkin (2004) reported, visual artists had used art forms to celebrate as well as critique genetics since the late 1980s in such works as portraits of DNA and apocalyptic paintings of genetically altered organisms. Taken after these, micrograph art offers opportunities for more socially and culturally sensitive representations and discussions of genetics.

But on the other hand, Nelkin's (1995) well-known critique that media and scientists rely on exaggeration to "sell" science rings true here as well. Sensational reports of scientific "breakthroughs" inspire public appreciation and admiration of science; these sentiments, in turn, lead to favorable research policies and public funding (Nelkin 1995). Micrographs, with their novel appearance on one hand and assumed nature as primary research evidence on the other, offer a perfect venue to "sell" science. It is thus prudent to ask whether social and ethical problems are being risked when these micrographs function more as "sales" strategies and rhetorical tropes than they offer contextualized understanding and deeper engagement.

This tendency is evident when popular science micrographs are stock images sourced from paid archives such as Science Photo Library and Science Source. This practice, judging by this study, is growing. Figures 3.9, 3.12 and 3.13, for example, are all stock images. Although chosen to be at least marginally relevant to a topic at hand, these images are, by definition, not created for the research context or exigency in question and, as such, are limited in their effect to communicate and engage. Furthermore, attempts toward striking appearance may result in the liberal use of color that, as mentioned earlier, is not accessible to all members of the public. More generally, it is conceivable that scientists and science communicators, if they use micrographs as a promotional attraction, may forgo less-than-striking visual evidence or omit markings and labels that disrupt readers' interpretation of beauty, even when such devices facilitate comprehension.

Conclusion

Modern genetic research's focus on cellular and molecular phenomena affords micrographs unique value as research evidence. In popular communication, this genre provides opportunities for publics to examine and assess primary research findings. But when micrographs are not made accessible syntactically, semantically, or pragmatically, they become not so much visual evidence as visual assurance that proper observation was made and conclusions were drawn, by someone. Reinforced, then, is the perception that genetics (and science in general) is an esoteric endeavor for the experts and that primary scientific evidence is too difficult for "lay" persons to interpret. Moreover, when micrographs look fantastic without being meaningful, their primary function becomes that of visual decoration, which contributes to readers' emotional interest but not cognitive interest or overall engagement. They are also, inevitably, bound up with science's social agenda for self-promotion.

At the same time, it is difficult to categorically deny the value of encouraging publics' emotional interest in and appreciation of science. Genetics deals with fascinating topics in nature and life, and if micrographs can help present its study in equally fascinating images, why not? These images can be framed to tell meaningful stories about research findings; they can help change the stereotypical view of science as dry and uninteresting; and more broadly, they serve to bridge arts/humanities and sciences so as to, in a more fundamental way, generate public participation and engagement in science.

But to realize these potentials, we need more targeted research. While research in microscopy is robust, this research focuses not on the presentation of micrographs but their procurement: for example, sample preparation, optical techniques, and tools and equipment (see Bozzola and Russell 1998; North 2006; Wiley's *Microscopy Research and Technique* journal). The assumption seems that what matters is the technical obtainment of data and evidence; as for the communication of that evidence, that is either commonsensical or secondary. The few guidelines one can find on micrograph presentation come from science journals' "instructions for authors" and textbooks that prepare authors to publish in such journals (e.g., Day and Gastel 2006). These guidelines go little beyond the short syntactic advice of cropping micrographs and using arrows and other markers for emphasis, adding scale bars to denote the degree of magnification, and considering the need of "colorblind" readers.[6] Semantic and pragmatic issues are overlooked. Moreover, because these guidelines are intended to help scientists prepare micrographs for fellow scientists, not all of them benefit public readers. Scale bars, for example, only benefit an audience who has the formal knowledge and experience to correlate, say, 100 nm with a half-inch bar and can use that correlation to envision the scale of an image.

What we need, then, is synergized research from natural scientists, science communication scholars, and visual communication scholars, research that prioritizes the challenges and opportunities associated with presenting micrographs for public readers. Such research should include rhetorical analyses of micrographs from different science disciplines, reader response studies based on select images, and ethnographic studies of scientists' and communicators' practices creating micrographs. It is with such work that we may hope to meaningfully complicate and contemplate the use of micrographs in popular science communication. As a genre that promises to, on one hand, depict microscopic scientific observation and, on the other, attract and engage readers, micrographs deserve no less.

Notes

1. According to the opponent process theory proposed by German psychologist Ewald Hering, red and green are opponent colors. That is, when cone cells that respond to red are activated, those that respond to green are inhibited, giving rise to our heightened perception of the two colors' contrast.
2. The researchers who created the micrograph had referred to the color as turquoise (greenish-blue), though it appears rather close to green and was, indeed, described as green in the *Science News* article that reported the research and reprinted the image.
3. The question of what consistutes a story or narrative is subject to different interpretations among narrative scholars (see e.g., Altman 2008). Accounting for all these interpretations is beyond the scope of this book.
4. I am indebted to conversations with Linda Duke, museum director and art scholar, for this idea.
5. In some of these cases, what is fearsome and splendid becomes a question of individual perception.
6. As seen in this study, such suggestion has not gone very far.

References

A Plague on All Our Houses. (1996, January). *Popular Science,* 50–56.

Altman, R. (2008). *A theory of narrative.* New York: Columbia University Press.

Anker, S., & Nelkin, D. (2004). *The molecular gaze: Art in the genetic age.* Cold Spring Harbor: Cold Spring Harbor Laboratory Press.

Antonarakis, S. E., Lyle, R., Dermitzakis, E. T., Reymond, A., & Deutsch, S. (2004). Chromosome 21 and Down syndrome: From genomics to pathophysiology. *Nature Reviews Genetics, 5*(10), 725–738.

Bakalar, N. (2006). Older paternal age seen as factor in some birth defects. *The New York Times.* Retrieved December 20, 2016, from http://www.nytimes.com/2006/06/06/health/06sper.html?.

Barthes, R. (1957). *Mythologies.* (Jonathan Cape Ltd., Trans.). New York: Noonday Press.

Beil, L. (2012, July). Catching a cancer: Viral culprits may explain a host of tumors with as-yet unknown triggers. *Science News, 182*(2), 22–25.

Bozzola, J. J., & Russell, L. D. (1998). *Electron microscopy* (2nd ed.). Sudbury, MA: Jones & Bartlett Publishers.

Brooks, P. (1992). *Reading for the plot: Design and intention in narrative.* Cambridge, MA: Harvard University Press.

Carretié, L., Hinojosa, J. A., Mercado, F., & Tapia, M. (2005). Cortical response to subjectively unconscious danger. *Neuroimage, 24*(3), 615–623.

Cheng, S. (2003, March 09). Her enigmatic science. *Los Angeles Times.* Retrieved July 21, 2017, from http://articles.latimes.com/2003/mar/09/entertainment/ca-cheng9.

Cohen, E. P., De Zoeten, E. F., & Schatzman, M. (1999). DNA vaccines as cancer treatment. *American Scientist, 87*(4), 328–335.

Cohn, N., Cohn, M., Paczynski, R., Jackendoff, P., Holcomb, G., & Kuperberg. (2012). (Pea)nuts and bolts of visual narrative: Structure and meaning in sequential image comprehension. *Cognitive Psychology, 65*(1), 1–38. doi:10.1016/j.cogpsych.2012.01.003.

Collins, F. (2014). Using genomics to follow the path of Ebola. *NIH Director's Blog.* Retrieved May 20, 2015, from http://directorsblog.nih.gov/2014/09/02/using-genomics-to-follow-the-path-of-ebola/.

Color Universal Design Organization. (2006). *Color universal design handbook.* Retrieved February 3, 2015, from http://www.eizo-apac.com/products/flexscan/color_vision/handbook.pdf.

Connelly, F. M., & Clandinin, D. J. (1990). Stories of experience and narrative inquiry. *Educational Researcher, 19*(5), 2–14.

Cornelis, J. F. V., Meade-Tollin, L. C., & Bosman, F. T. (1998). Metastasis. *American Scientist, 86*(2), 130–141.

Dahm, R. (2006). The zebrafish exposed. *American Scientist, 94*(5), 446–453.

Day, R., & Gastel, B. (2006). *How to write and publish a scientific paper* (6th ed.). Westport, CT: Greenwood Press.

de Duve, C. (1995). The beginnings of life on earth. *American Scientist, 83*(5), 428–437.

Dinolfo, J., Heifferon, B., & Temesvari, L. A. (2007). Seeing cells: Teaching the visual/verbal rhetoric of biology. *Journal of Technical Writing and Communication, 37*(4), 395–417.

Dragga, S. (1992). Evaluating pictorial illustrations. *Technical Communication Quarterly, 1*(2), 47–62. doi:10.1080/10572259209359498.

Ede, S. (2002). Science and the contemporary visual arts. *Public Understanding of Science, 11*(1), 65–78. doi:10.1088/0963-6625/11/1/304.

Edwards, J. L. (1997). *Political cartoons in the 1988 presidential campaign: Image, metaphor, and narrative.* New York: Garland Publishing.

Epstein C. J. & Golbus, M. S. (1977). Prenatal diagnosis of genetic diseases. *American Scientist, 65*(6), 703–711.

Fischhoff, B. (2013). The sciences of science communication. In *Proceedings of the National Academy of Sciences of the United States of America, 110*, (pp. 14033–14039). doi:10.1073/pnas.1213273110.

Ford, B. J. (1993). *Images of science: A history of scientific illustration.* New York: Oxford University Press.

Gage, F. H., & Muotri, A. R. (2012, March). What makes each brain unique. *Scientific American*, 26–31.

Galison, P. (1998). Judgment against objectivity. In C. A. Jones, P. Galison, & A. E. Slaton (Eds.), *Picturing science, producing art* (pp. 327–359). New York: Routledge.

Gallagher, J. (2013). Y chromosome practically obliterated in mice. *BBC News.* Retrieved May 20, 2015, from http://www.bbc.com/news/health-24991843.

Gaviria, A. R. (2008). When is information visualization art? Determining the critical criteria. *Leonardo, 41*(5), 479–482.

Gibbs, W. W. (2003). Untangling the roots of cancer. *Scientific American*, 56–65.

Goldsmith, E. (1984). *Research into illustration: An approach and a review.* Cambridge: Cambridge University Press.

Gross, A. G., & Harmon, J. E. (2013). *Science from sight to insight: How scientists illustrate meaning.* Chicago: University of Chicago Press.

Guillette, L., Gross, T., Masson, G., Matter, J., Percival, H., & Woodward, A. (1994). Developmental abnormalities of the gonad and abnormal hormone concentrations in juvenile alligators from contaminated and control lakes in Florida. *Environmental Health Perspectives, 102*(8), 680–688.

Hoffman, M. (1994). AIDS. *American Scientist, 82*(2), 171–177.

Hooke, R. (1665). *Micrographia: Or some physiological descriptions of minute bodies made by magnifying glasses with observations and inquiries thereupon.* London: The Royal Society.

Jameson, D. A. (2000). Telling the investment story: A narrative analysis of shareholder reports. *Journal of Business Communication, 37*(1), 7–38. doi:10.1177/002194360003700101.

Jegalian, K., & Lahn, B. (2001, February). Why the Y is so weird. *Scientific American, 284*(2), 56–61.

Karidis, A. (2015, March 9). Doctors study tumors' genetic makeup to guide cancer treatment. Retrieved December 16, 2015, from https:// www.washingtonpost.com/national/health-science/doctors-study-tumors-genetic-makeup-to-guide-cancer-treatment/2015/03/09/d1e3cc78-a0e7-11e4-9f89-561284a573f8_story.html.

Kiesselbach, T. A. (1951). A half-century of corn research. *American Scientist, 39*(4), 629–655.

Kim, H., Sheffield, D., & Almutairi, T. (2014). Effects of fear appeals on communicating potential health risks of unregulated dietary supplements to college students. *American Journal of Health Education, 45*(5), 308–315.

Kjærgaard, R. S. (2011). Things to see and do: How scientific images work. In D. J. Bennett & R. C. Jennings (Eds.), *Successful science communication: Telling it like it is* (pp. 332–354). Cambridge: Cambridge University Press.

Koch, R. (1877). Verfahren zur untersuchung, zum conservieren und photographiren der bakterien. *Beiträge zur Biologie der Pflanzen, 2,* 399–434.

Lane, M. A. (1908, July 18). Weismannism. *Scientific American Supplement,* 45–46.

LaTour, M. S., Snipes, R. L., & Bliss, S. J. (1996). Don't be afraid to use fear appeals: An experimental study. *Journal of Advertising Research, 36*(2), 59–67.

Lynn, F. M. (1990). Public participation in risk management decisions: The right to define, the right to know, and the right to act. *Risk, 1,* 95–101.

Macdonald, S. (1996). Authorising science: Public understanding of science in museums. In A. Irwin & B. Wynne (Eds.), *Misunderstanding science? The public reconstruction of science and technology* (pp. 84–106). Cambridge: Cambridge University Press.

McGhee, J. (2010). 3-D visualization and animation technologies in anatomical imaging. *Journal of Anatomy, 216*(2), 264–270.

McLachlan, J. A., & Arnold, S. F. (1996). Environmental estrogens. *American Scientist, 84*(5), 452–461.

Miller, O. (1973). The visualization of genes in action. *Scientific American, 228*(3), 34–42.

Miller, O., Hamkalo, B., & Thomas, C. (1970). Visualization of bacterial genes in action. *Science, 169*(3943), 392–395.

Mone, G. (2003, June). The ultimate exterminator. *Popular Science,* 41–42.

National Cancer Institute. (2001). Karyotype (normal). Retrieved August, 2, 2016, from https://visualsonline.cancer.gov/details.cfm?imageid=2721.

National Media Museum. (2016). Microphotograph of a flea. Retrieved July 13, 2016, from http://www.nationalmediamuseum.org.uk/collection/photography/royalphotographicsociety/collectionitem?id=2003-5001/2/23760/2.

National Science Foundation. (2015a). 2011 Winners. Retrieved May 21, 2015, from https://www.nsf.gov/news/special_reports/scivis/winners_2011.jsp.

National Science Foundation. (2015b). *The Vizzies Visualization Challenge.* Retrieved May 21, 2015, from https://www.nsf.gov/news/special_reports/scivis/challenge.jsp.

Nelkin, D. (1995). *Selling science: How the press covers science and technology.* New York: W.H. Freeman.

Nettelbeck, D. M., & Curiel, D. T. (2003, October). Tumor-bursting viruses. *Scientific American. 289*(4), 68–75.

North, A. J. (2006). Seeing is believing? A beginners' guide to practical pitfalls in image acquisition. *The Journal of Cell Biology, 172*(1), 9–18.

Okabe, M., & Ito, K. (2008). Color University Design (CUD): How to make figures and presentations that are friendly to colorblind people. Retrieved May 20, 2015, from http://jfly.iam.u-tokyo.ac.jp/color/#checker.

Overney, N., & Overney, G. (2011). *The history of photomicrography.* Retrieved May 20, 2015, from http://www.microscopy-uk.org.uk/mag/artmar10/history_photomicrography_ed3.pdf.

Patterson, D. (1987). The causes of Down Syndrome. *Scientific American,* 257(2), 52–60.

Perkins, J., & Blyler, N. (1999). Introduction: Taking a narrative turn in professional communication. In J. Perkins & N. Blyler (Eds.), *Narrative and Professional Communication* (pp. 1–34). Stamford, CT: Ablex.

Price, F. (1996). Now you see it, now you don't: Mediating science and managing uncertainty in reproductive medicine. In A. Irwin & B. Wynne (Eds.), *Misunderstanding science? The public reconstruction of science and technology* (pp. 84–106). Cambridge: Cambridge University Press.

Reichert, T., & Lambiase, J. (Eds.). (2005). *Sex in consumer culture: The erotic content of media and marketing.* London: Routledge.

Ruckes, H., & Mok, M. (1931, November). Secretes of sex explained by science. *Popular Science,* 38–40, 130–132.

Sachs, J. S. (2008, February). This germ could save your life. *Popular Science,* 64–69, 90, 92, 94.

Saey, T. H. (2009, April). Shared differences: The architecture of our genomes is anything but basic. *Science News, 175*(9), 16–20.

Saey, T. H. (2010, July). All patterns great and small: Researchers uncover the origins of creatures' stripes and spots. *Science News,* 28–29.

Schimke, R. T. (1980). Gene amplification and drug resistance. *Scientific American,* 60–69.

Snipes, R. L., LaTour, M. S., & Bliss, S. J. (1999). A model of the effects of self-efficacy on the perceived ethicality and performance of appeals in advertising. *Journal of Business Ethics, 19*(3), 273–285. doi:10.2307/25074096.

Stemmer, W., & Holland, B. (2003). Survival of the fittest molecule. *American Scientist, 91*(6), 526–533.

Stone, M. (2009). Information visualization: Challenge for the humanities. *Working together or apart: Promoting the next of digital scholarship* (pp. 43–56). Washington, D.C.: Council on Library and Information Resources.

Thone, F. (1927). X-rays speed up evolution over 1,000 per cent. *The Science News-Letter, 12*(340), 243–246.

Travis, J. (1995). End games. *Science News, 148*(22), 362–363, 365.

Travis, J. (1997). Eye-opening gene. *Science News, 151*(19), 288–289.

Trolio, J. (2016, February). FYI. *Popular Science,* 87–88.

van Dijck, J. (1998). *Imagenation: Popular images of genetics.* New York: New York University Press.

Ware, C. (2012). *Information visualization: Perception for design* (3rd ed.). Burlington, MA: Morgan Kaufmann.

Weinschenk, S. (2011). *100 things every designer needs to know about people.* Upper Saddle River, NJ: Pearson Education.

Werner, T., Koshikawa, S., Williams, T. M., & Carroll, S. B. (2010). Generation of a novel wing colour pattern by the Wingless morphogen. *Nature, 464*(7292), 1143–1149.

Wilson, E. B. (1913, August 2). Heredity and Microscopic Research. *Scientific American Supplement,* 77–80.

Wong, B. (2011). Color blindness. *Nature Methods, 8*(6), 441.

4

The Illustrated View: Iconic Explanations and Figurative Metaphors

Scientific illustrations are as old as science itself and a dominant form of visual communication used across different branches of science (see Ford 1993). As a visual genre, it conceptualizes scientific concepts and conveys experimental processes, which are fundamental to all scientific inquiries. Accordingly, in the public communication of genetics, illustrations accompanied the rise of classical genetics and remained prevalent throughout the decades despite changed research focus and approach.

In Galison's (1998) historical study of scientific visuals, illustrations epitomize the latest visual paradigm of interpreted images. That is, this visual genre is not interested in recording nature "as is" but instead acknowledges, indeed, celebrates, expert "judgment" and selective filtering of evidence. By filtering reality and reducing complex reality to its salient features, illustrations are believed to be accessible to non-expert audiences. However, one needs to consider that such accessibility is subject to how illustrations parse visual information and whether viewers are familiar with the kinds of visual "codes" used in the resultant illustration (Amare and Manning 2013). Moreover, as interpreted images, illustrations provide creators with ample freedom for syntactic choices, choices that are influenced by social–cultural contexts and can produce

© The Author(s) 2017
H. Yu, *Communicating Genetics*,
DOI 10.1057/978-1-137-58779-4_4

distinct, if sometimes subtle, semiotic effects. In particular, as this chapter details, the illustrations of DNA, the household name for genetics, carry rich extra-scientific implications.

It should be noted that the term *illustration* is often selectively defined by different writers. The concept easily converges with related terms such as *drawing, schematic,* and *diagram* and may be defined to be either a higher or lower category of such terms (see, e.g., Amare and Manning 2013). This chapter does not attempt to offer a general definition for *illustration,* as that is beside its point; rather, it uses the term in a localized context to refer to hand-drawn and computer-generated images that employ syntactic elements such as lines, shapes, and colors to depict genetics objects/phenomena or experiment procedures.

Minimalist Illustrations: Reduced Accessibility and Enhanced Ethos

The era of classical genetics, as mentioned in previous chapters, focuses on the physical observation of inheritance and mutation. Though such observation is conveniently and convincingly demonstrated via photographic records (see Chap. 2), illustrative representations also exist. As an interpretation of the observed reality, the illustrated view produces different cognitive and affective effects than photographs of the same subject.

Consider Fig. 4.1, which illustrates the process of crossbreeding two strains of Phlox flower. In the parent generation (denoted as P), a flat, salver-shaped flower (left) is cross-pollinated with a funnel-shaped flower (right). Their first-generation (F1) flowers are all salver-shaped. But when two F1 flowers are crossed, their offspring (F2) include both salver- and funnel-shaped flowers. These patterns, as the illustration shows, can be explained via the inheritance of dominant and recessive gene forms (i.e., alleles). The salver-shaped parent carries two copies of the dominant allele (SS); the funnel-shaped parent carries two copies of the recessive allele (ss). When crossed, each parent randomly contributes one allele to the offspring, creating an entire F1 generation of Ss; with S being dominant, all F1 flowers are salver-shaped. When two F1 flowers (Ss and Ss) are crossed, the same random contribution results in SS, Ss,

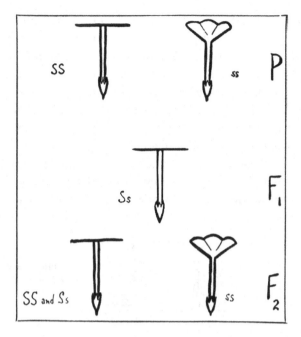

Fig. 4.1 Simple and concrete illustration of plant heredity (Kelly 1922, p. 177)

and ss. With S being dominant, SS and Ss offspring are salver-shaped, and ss offspring are funnel-shaped.

Figure 4.1 represents the physiology of Phlox flowers via simple geometric shapes: a flat line plus a vertical column or a funnel shape plus the vertical column. Imitating reality via physical likeness, these images are what Peirce (1894) classified as icons. By reducing complex reality to essential features, icons are generally acknowledged to enhance readers' comprehension and recall of information (McCloud 1994; Shirk and Smith 1994; Katz et al. 2006). In particular, the icons used in Fig. 4.1 are low in complexity; that is, only a small amount of visual detail is physically embodied in the icons (McDougall et al. 2000). Lower complexity, as studies suggest, facilitates icon recognition and visual search (Byrne 1993; McDougall et al. 2000). In this case, it allows the illustration to discard all details of the flower that may distract readers and guide readers to focus on the one and only topic of interest: its shape.

Such iconic choices are typical of pre-1980s illustrations used in the popular communication of genetics. Certainly, in saying this, I do not suggest that all early illustrations contain the same extreme minimalism seen in Fig. 4.1; a range of possibilities do exist. Figure 4.2, which also diagrams Mendelian cross-generation inheritance/mutation, uses icons with relatively higher complexity. As the figure explains, when a fruit fly exposed to X-ray mates with a normal fruit fly, their first-generation offspring appear normal, but when the first generation flies are bred, their offspring develop various mutations, including underdeveloped or undeveloped wings and abnormal eye colors. Compared with Fig. 4.1, Fig. 4.2 uses more detailed line work to portray the physical structure

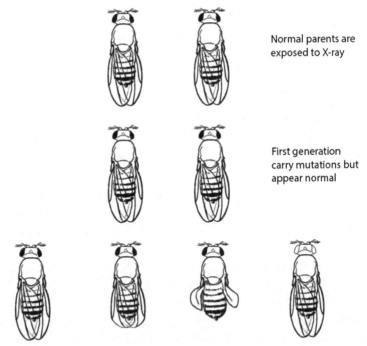

Normal parents are exposed to X-ray

First generation carry mutations but appear normal

Among the second generation, some are normal (left), some have underdeveloped wings (2nd from left) or undeveloped wings (3rd from left), and some have white eyes (right)

Fig. 4.2 Moderately complex but concrete illustration of fruit fly mutations. Recreated based on Thone (1927, p. 244)

of its specimen. The color black is used to not only emphasize eye mutations but also depict the flies' segmented abdomen.

Despite these syntactic details, the icons used in Fig. 4.2 are still on the low end of the complexity spectrum as compared with later illustrations (more about this below). Moreover, its modest increase in complexity, from an iconic standpoint, does not render the figure more difficult to recognize or understand than Fig. 4.1. This is because both figures use icons that are high in concreteness. In icon studies, concreteness is a characteristic that measures the extent to which icons represent things that viewers are familiar with in the real world; when concreteness is high, viewers can apply what they already know about those things to more easily recognize the icons (McDougall et al. 2000, p. 292). Figures 4.1 and 4.2, in depicting common or at least relatable objects, are both high in concreteness.

This situation changed when genetics transitioned from the classical era to the DNA age in the 1950s. In the next three decades, tremendous progress was made in genetic research: The DNA double-helix structure was resolved, the mechanism of gene expression was determined, and DNA recombination, amplification, and sequencing techniques were established. These and other advancements gave rise to an era of illustrations that portray abstract and complex cellular and molecular objects and experiment procedures.

Figure 4.3 is one such example. In this case, the image illustrates the process of herpes virus infection: the virus enters a healthy human cell, replicates viral DNA, hijacks the host's protein translation mechanism to make viral proteins, assembles new viruses, and finally destroys the cell. The image, as is common with illustrations from this period, is semantically complex, involving multitudes of entities and steps and entailing lengthy captions. Its individual icons, on the other hand, remain low in complexity: Plain circles and triangles represent viruses and cells; simple lines, rather than the default double-helix structure seen in present-day portrayals, denote DNA. Color, which started to appear during this time, is used sparingly: Only red and black are used, and their use is limited to information conveyance, that is, color coding different visual elements or highlighting an element of interest.

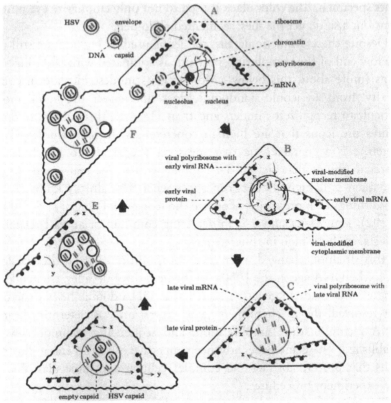

Figure 3. Schematic representation of 6 stages of herpes simplex virus replication in a human cell. A. Adsorption of the virion to the uninfected cell is followed by uncoating of the intact virion and entry of viral DNA into the host nucleus. B. Early viral mRNA is transcribed and translated into early viral proteins, some of which alter host-cell membranes and functions, and some of which are necessary for replication of viral DNA. C. Viral DNA is replicated, and both early and late viral mRNA are transcribed and translated into viral proteins. D. Host-cell membranes are further affected by viral proteins, and viral structural proteins encapsulate newly synthesized DNA in the nucleus to form HSV capsids. E. There is a further decrease in biosynthetic activity of the infected cell. The nuclear membrane becomes extremely convoluted and begins to envelope viral capsids. F. Almost all biosynthetic activity has now ceased. Many enveloped virions are found in the cytoplasm; some are released from the cell, and these, or the infected cell itself, can now fuse with an uninfected cell to start a new round of replication.

Fig. 4.3 Semantically complex but iconically minimal illustration of herpes virus infection (Wagner 1974, p. 587). Reprinted by permission of *American Scientist*, magazine of Sigma Xi, The Scientific Research Society

Illustrations like Fig. 4.3, in their effort to outline research findings and biological processes in uttermost detail, align with the kind of public science communication that was prevalent at the time, what was later described the deficit model (more about the deficit model see Chap. 1).

According to this model, scientists need to make effort to carefully profess their work to the public, and the public needs to make effort to absorb those formal, proven findings; when both do, the result is an enlightened public who supports and is enthusiastic about science. Under such assumptions, illustrations should, as Fig. 4.3, focus on getting the science "right" with austere details and minimal flare and distraction.

Retrospectively, it is easier to see the problem with such assumptions. Compared with classical genetics illustrations such as Figs. 4.1 and 4.2, illustrations from the 1950s–1980s are notably less accessible, both cognitively and affectively. Granted, Fig. 4.3 and similar illustrations from its era (see, e.g., Stein et al. 1975; Yuan and Hamilton 1982) retain the low iconic complexity seen in earlier works, which should facilitate icon recognition and visual search. However, in these cases, this benefit is overshadowed by the icons' lack of concreteness. Whether they are DNA, RNA, genes, proteins, or similar entities dealt with in the DNA age, they are usually invisible and unfamiliar to most public readers. As such, readers will find these iconic representations lacking concrete reference and thus difficult to recognize or understand.

Related to this factor is the amount of background knowledge these illustrations expect from readers. In the case of Fig. 4.3, one needs to at least be somewhat familiar with protein translation and the multiple entities involved in that process to readily understand the illustrated steps. These illustrations' liberal use of arrows also belies this assumed knowledge[1]: Readers are asked to know when arrows mean physical movement, when they indicate the progression of procedural steps, when they refer to the increase/decrease of measurements, and when they denote the addition/removal of elements. In Fig. 4.3, while the bold arrows between the alphabetized steps are relatively straightforward in signifying the progression of steps, the multitude of others within the image are less so, unless readers already know what they are looking at.

It is true that background knowledge *can* be given in an accompanying caption, but large amounts of such knowledge often cannot be effectively conveyed when the caption must account for the illustrated elements and be reasonably concise. Figure 4.3's original caption provides some background but omits details such as the role of the depicted polyribosome. Still, longer captions will not be of much help either. As studies in cognitive psychology show, an individual's working memory or processing capacity is inherently limited (Miller 1994; Ware 2012). When large numbers of visual and textual elements are separately presented, as in the case of a complex illustration and a long caption, readers have to continuously glance back and forth to integrate disparate information, which taxes the working memory and reduces the image's accessibility (Sweller 1994).

In short, from classical genetics to the 1950s–1980s DNA age, scientists and science communicators' strategy to employ minimalist icons persisted, but that strategy was losing its ability to create accessible illustrations because of the increase in the knowledge base, research complexity, and visual abstraction. Significantly, as these illustrations' accessibility decreased, their projected ethos as scientific evidence was confirmed and enhanced, and with it, genetics' social status as a respected discipline of modern science.

By using minimalist lines and shapes, illustrations, as early as those from the classical genetics era, create a sense of clinical detachment and parsimonious analysis. Compared with photographic depictions of Mendelian crossbreeding (see Chap. 2), illustrations such as Figs. 4.1 and 4.2 no longer convey practical outcomes. Instead, they say, as it were, what is represented here is "not a means of looking *at* the world, but a means of looking *through* it to its causal structure" (Gross and Harmon 2013, p. 81). In other words, these images represent the ethos of modern science as that which objectively and disinterestedly describes and explains nature. It is interesting to note that such a visual impression effectively masks the nature of illustrations as interpreted images. A scientist's version of interpretation, when visualized through minimalist and abstract icons, becomes *the* version of reality.

This ethos of objectivity and authority was enhanced in the 1950s–1980s when the use of unadorned lines and shapes multiplied in illustrations. What is also added is a feeling of unfamiliarity and intimidation, due to illustrations' increased semantic detail and reduced iconic concreteness. If illustrations from the classical era created an appearance of genetics as an emerging (and humble) discipline closely tied to agriculture and extension work, these later works marked the coming of age of genetics as the new scientific frontier with abstract (and thus far-reaching) implications. Not only *are* these illustrations more difficult to comprehend for a non-expert, they also *look like* they would be rather difficult to comprehend—the ultimate ethos of specialized research. While this new ethos may command respect and appreciation from fellow scientists and even "fans" of science, for others, it gives rise to a perception of genetics as an esoteric endeavor removed from everyday readers and their concerns.

Hyper-Realistic Illustrations: Attempts for Accessibility and Engagement

Toward the end of the twentieth and beginning of the twenty-first century, when genetics was pushed into the center of public attention as the epitome of medical research, the landscape of popular science illustrations changed again. Shedding the minimalist look, these contemporary images favor complex icons with rich syntactic details such as shading, shadow, texture, lightness contrast, color, and perspective drawing (see, e.g., Nettelbeck and Curiel 2003; Saey 2008; Greenwood 2013). Figure 4.4, which is used in a *Scientific American* article (Misteli 2011, p. 70) to illustrate the process of gene activation, is an example. As the figure explains, gene activation starts with transcription factors (a type of protein) collecting on the DNA's regulatory region. This action enables RNA polymerase (an enzyme) to transcribe DNA into messenger RNA, which is further translated into protein by the ribosome (a tiny organelle that exists in large quantities in living cells).

Compared with earlier illustrations, Fig. 4.4 uses icons with much higher visual complexity. The DNA segment appears, as it is customary today, in a colored double helix. Different parts of the RNA polymerase assume dark, medium, and light shades of blue, creating varying light reflections and an illusion of depth. Detailed shadowing and highlighting work is similarly used on the transcription factors and creates a sense of uneven texture. All objects are drawn with two-point perspective,[2] a common technique to create the appearance of three dimensions.

These syntactic details, according to conventional, cognition-based principles of information design, are extra ink that does not add essential visual information (Tufte 2001; Malamed 2009). Because of this, the illustration is said to have a high "ink-to-data" ratio, which means readers have to search through more perceptual clues for information; the demand on readers' working memory is thus increased, and information efficiency and effectiveness are said to decrease.

On the other hand, if we go beyond cognitive processing and consider an image's affective impression or overall semiotic effect, Fig. 4.4 and similar images may well have an advantage for public engagement, especially when compared with the intimidating illustrations from the 1950s–1980s. To start, Fig. 4.4 uses color much more liberally. Color is no longer restricted for information conveyance but also visual appeal and hence, emotional interest. In this case, pastel colors (soft green, yellow, blue, etc.) adorn the various organelles and make the illustration

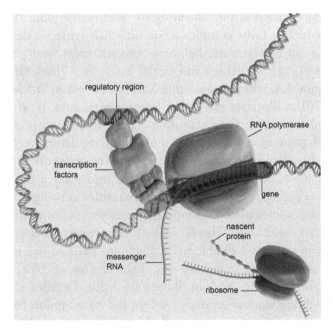

Fig. 4.4 Gene activation depicted with rich syntactic details. Image courtesy of James Archer/Anatomy Blue (color figure online)

appear pleasant and welcoming. Contributing to the same impression, the organelles appear in soft angles rather than defined geometric shapes. More importantly, these objects, much like objects in everyday life, have realistic details such as texture and dimension, which make these otherwise unfamiliar cellular and molecular components *appear as if* easily recognizable or relatable. Meanwhile, the illustration still looks intriguing because its depicted objects, complete with vivid details, are not *quite* like what one observes in everyday life.

These perceptual cues are not inconsequential: according to psychology and cognitive studies, when a learning environment presents such features as novelty and comprehensibility, it can arouse interest and help readers to selectively focus, to use prior knowledge, to maintain higher levels of engagement, and to comprehend and recall information (Schraw 1998; Park and Lim 2007). Similarly, as Christiansen (2013) argued from an information design perspective, while minimalist and abstract illustrations may be well suited to communicating complex ideas with specialist audiences, they can be "off-putting" when used with public readers. "Beautifully detailed and somewhat fancifully figurative art" can better capture non-specialist readers' imagination and attention, offer them an entry point to the science in question, and serve, in general, to engage and inspire (Christiasen 2013, p. 49).

In a way, then, these newest illustrations align with contemporary public science communication's emphasis on personal relevance and broad engagement. In purposefully seeking visual attraction, they also acknowledge the reality of today's popular science communication, that is, as one form of mass media products that must compete with other leisure channels for reader attention. As seen in earlier chapters, photographs and micrographs have their fair share of creating eye-catching images, yet illustrations, as interpreted images and by definition a product of individual stylistic choices, afford more creative license.

As some readers may suspect, these latest stylistic tendencies are not without potential drawbacks. Texts and images that are highly interesting but only tangentially relevant have been referred to as "seductive details" by educational psychologists. Such details, some scholars argue, distract readers away from core information, disrupt information processing, or prime readers to activate irrelevant knowledge schemas to organize information (Park and Lim 2007; Harp and Mayer 1997, 1998).[3] In terms of scientific visual representations, these attractive details are not merely tangential (and thus "still good to have"); they are semiotically excessive (and thus potentially misleading). That is, cellular and molecular objects defy common ways of "seeing." Their structures are often deduced through X-ray crystallography and NMR spectroscopy[4] and rendered in terms of atomic structures. An RNA polymerase, for example, rather than the solid, textured object seen in Fig. 4.4, has an atomic structure as depicted in Fig. 4.5.

By this, I am not suggesting that Fig. 4.5 is *the* right way to depict an RNA polymerase or other cellular and molecular objects; what it is is a structural representation and one possibility at that (more about structural representations see Chap. 7). Rather, my point is to highlight the extent of creative license entertained in contemporary illustrations. The issue with these added details is not, as some may say, one of reduced "accuracy"—minimalist illustrations that use simple icons to represent organelles are no more accurate than embellished illustrations. The issue, rather, lies in the false sense of concreteness and realism

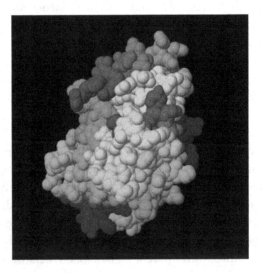

Fig. 4.5 A space-filling model view of human RNA polymerase II (Image from the RCSB PDB of PDB ID 2C35 2005) (color figure online)

projected by hyper-detailed illustrations. As Molyneaux (2013) put it, the "use of naturalistic imagery implies a direct relationship between the representation and the world, transparent and without interpretive obstacles. The ideas represented claim truth to nature" (p. 2). While minimalist icons cue readers of abstract concepts, icons with vivid and commonly perceived visual details suggest cellular and molecular reality.

This perceptual difference can be explained via McCloud's (1994) iconic face spectrum. Given the two faces in Fig. 4.6, most viewers would not readily associate the left face with a real person, but many would do so with the right face, not because they know or have seen this person but because the face's syntactic richness makes it *appear* concrete and real. Because of this effect, hyper-realistic genetics illustrations may be said to conflate artistic invention and physical reality. They suggest that genetics *is* about these concrete, if somewhat fanciful, objects; that cellular and molecular organelles *are* like everyday objects; and that scientific investigation *should* therefore be familiar, engaging, and even fun. To borrow Macdonald's (2004) words, such "strategies of familiarisation and accessibility may act as a kind of intellectual narcotic," leading viewers into "*not* asking questions, into a sense of security that the world of science *is* familiar" (p. 168).

Moreover, hyper-realistic icons, though they may seem "harmless" in a given image, can pose issues as part of readers' growing visual

Fig. 4.6 Icons with rich syntactic details suggest realism

repertoire. To start, because different visual creators use different syntactic techniques to depict the same object, readers may find that the rich, detailed visual knowledge they gained from one illustration does not necessarily transfer to the next. More significantly, some artistic inventions, as they try to evoke familiar visual imagery, end up creating barriers or at least ambiguities. Consider, for example, illustrations that use metal blades to depict some enzymes' ability to cleave DNA strands (June and Levine 2012) or illustrations that depict uncoiled chromosomes as color-coded balls of threads (Misteli 2011). Surely, readers will recognize these as artistic inventions—or will they?

Figure 4.7 is a case in point. It accompanies a National Institutes of Health news article about the Fragile X Syndrome, a disorder associated with the X chromosome, which is presumably illustrated in Fig. 4.7. The image is carefully embellished with color and lightness contrast: the chromosome has a shining outline; it floats on a dark, cosmic background;[5] and a double helix, which most readers will recognize as DNA, peers through the chromosome in a contrasting red color. Together, these syntactic details portray the chromosome as a transparent, hollow object that houses DNA inside it. Vividly realistic, the image is a concrete reference to two basic genetics concepts—only it is rather misleading, as chromosomes do not enclose DNA but *are* highly condensed DNA and proteins.

By these critiques, I am not suggesting that we abandon all visual embellishments in popular science genetics illustrations. As discussed

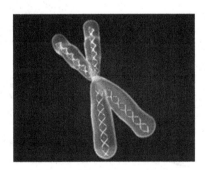

Fig. 4.7 Chromosome as hollow object that houses DNA (National Institutes of Health 2009) (color figure online)

earlier, syntactic details *are* affectively relevant and help create audience engagement. What I do intend to do is calling attention to the range of issues involved in hyper-realistic illustrations. Although these illustrations are apparently taken for granted in contemporary popular science communication, their full implications are often ignored in the literature. When we celebrate these images' active consideration of public readers, their ability to attract attention, to cultivate a sense of accessibility, it is also useful to consider what these newest illustrations may have to lose for their gains.

At the very least, visual creators should consider illustrations' cognitive function as well as affective effect so their work conveys meaningful information *and* remains visually welcoming. Regarding Fig. 4.7's topic of chromosomes, for example, syntactic details can be used to create iconic concreteness and a sense of accessibility without being excessive or misleading. Figure 4.8, which explains the phenomenon of chromosome deletion, is a possible example of how to do so. In Fig. 4.8, colors and shading are used to render chromosomes as solid "bars" with a wiry surface. The chromosomes look relatable, if not familiar; indeed, they take on a "lumpy" texture that is not unlike how they appear under the microscope. Meanwhile, the process of chromosome deletion is illustrated with minimal effort via the help of colors and call-out labels.

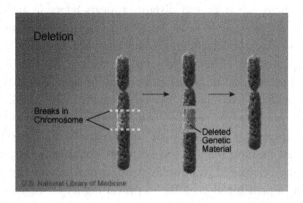

Fig. 4.8 Illustration of chromosome deletion that considers both cognitive and affective factors (Lister Hill National Center 2016, p. 26) (color figure online)

It is also useful to note that realistic details are not the only way to create visual appeal. In an opposite move from hyper-realistic, digitally enhanced illustrations, some (granted, very few) contemporary popular science genetics illustrations are purposefully simple, casual, and sketchy. This simplicity is different from the minimalist style seen in earlier decades. That is, although these recent illustrations similarly use simple lines and shapes, they do not look clinically detached as earlier illustrations do, as the example in Fig. 4.9 should demonstrate. This particular illustration compares the old and new ways of developing antibiotics. The old way relies on untargeted testing of extracts for signs of bacterial inhibition; promising extracts are then modified to improve stability or reduce toxicity. The new approach, by contrast, analyzes pathogens' genome sequences and uses sequence data as well as knowledge of protein structures to identify essential bacterial genes and proteins. Specific antibiotics that target these genes and proteins can then be developed.

In Fig. 4.9, lines are deliberately crooked, shapes are purposefully inconsistent, saturated colors remind one of a kid's drawing, and even the font resembles those found in a comic book. With these stylistic choices, the illustration appears distinctly cartoonish, disarming, and lighthearted. This visual effect, though unexpected or precisely because it is unexpected in science communication, makes the image seem easy and fun to peruse. In turn, the science behind the image, rather than being abstract and distant, appears relatable and accessible to any and all readers. It is as Yu (2015) argued, comics-style illustrations have notable advantages when used for technical and scientific communication: They help engage readers via fun caricatures; they encourage the use of familiar, everyday images; and they decenter experts' worldview to allow the participation of public readers. While it is possible, as Yu (2015) acknowledged, that some readers find such representations "childish" or irrelevant to "serious" scientific pursuits, making science less stilted without reducing its significance is to be celebrated in popular communication. At the very least, it represents a possible alternative to prevailing hyper-realistic popular science illustrations.

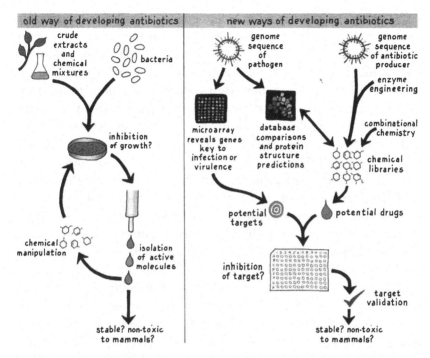

Fig. 4.9 Comics-style illustration compares old and new ways of developing antibiotics (Amábile-Cuevas 2003, p. 147). Reprinted by permission of *American Scientist*, magazine of Sigma Xi, The Scientific Research Society (color figure online)

Metaphorical Illustrations: The Case of the DNA Double Helix

Though recent popular science genetics illustrations are often high in iconic complexity and creative realism, their primary function remains to convey scientific information. In that sense, they are comparable to the informative photographs discussed in Chap. 2. By contrast, other illustrations from the last 20 or so years, as in the case of some photographs, ceased to function as scientific evidence or at least as any specific or operational evidence. Following the term used in Chap. 2, I refer to these illustrations as symbolic illustrations, illustrations whose primary purpose is to channel and foster viewer reactions and emotions rather than delineate scientific concepts or processes.

Compared with symbolic photographs, symbolic illustrations have a much more limited range of subjects in the communication of genetics. This is because for an image to take on symbolic connotations, it must first be something that readers are well familiar with and can readily recognize and interpret. Most cellular and molecular organelles and experiment processes depicted in genetics illustrations do not quite meet this requirement—with the exception of deoxyribonucleic acid, or DNA. As the "carrier" of genetic information and the substance of genes, DNA has a central, if not *the* central, place in the study of modern genetics since the 1950s. It is considered the biological determinants of organisms' individual traits, disease risk factors, and life functions. This importance was solidified by the completion of the Human Genome Project at the turn of the twenty-first century, which up-scaled DNA research from the level of single genes to complete genomes. DNA's biological importance, plus the fortuitous fact that its double-helix molecular structure is visually intriguing, makes DNA a prevalent symbol used by TV, movies, and advertising to suggest, often loosely, any and all things related to genetics, genetic engineering, biological experiments, and modern medicine. By repurposing this well-recognized symbol, popular science genetics illustrations are able to create a range of connotations.

Some such illustrations do so at the most basic level of visual appeal, that is, by using the DNA molecule as a syntactic decoration. The double-helix structure is, after all, an intricate design and one not commonly seen in the everyday environment. Its appearance, especially when infused with syntactic embellishments such as color and perspective drawing, serves to create visual attraction. For example, in Seeman (2004), a report on the use of DNA in nanodevices, long strands of colorful double helices are seen running down the pages, in much the same way a decorative border does.

More symbolically rich and semiotically interesting than these decorations are illustrations of DNA that invoke visual metaphors. In these instances, the double helix is employed to seemingly create accessible information but ultimately only affirm clichéd, ambiguous impressions.

Visual Metaphor

In text-based discourses, a metaphor is generally understood to be a figure of speech that allows readers to understand an unfamiliar or abstract concept (the target) by comparing it to a familiar or tangible concept (the source). The metaphor "life is a roller coaster," for example, equates the abstract idea of life having its ups and downs (the target) to the more concrete experience of riding a roller coaster (the source). Because of this ability to bridge the known and the unknown, metaphors are commonly used in science communication in general and popular science communication in particular (Giles 2008).

Beyond their role as a cognitive learning aid, metaphors are considered by some scholars to be a fundamental way of thinking and feeling about the world around us. George Lakoff and Mark Johnson (2003), in their seminal work *Metaphors We Live By*, argued that "metaphor is pervasive in everyday life, not just in language but in thought and action. Our ordinary conceptual system, in terms of which we both think and act, is fundamentally metaphorical in nature" (p. 3). That is, the select use of metaphors frames how we perceive and evaluate the people and things around us. The metaphor *argument is war* guides us to see the combative nature of argumentation: two parties trying to prevail over one another; however, invoking an alternative metaphor *argument is dance* allows us to see the cooperative aspect of argumentation: two parties trying to reach a mutual understanding (Lakoff and Johnson 2003).

In science, metaphors are pervasive ("evolution is natural selection" or "the immune system is life's defense mechanism"); these metaphors assume constructive power by invoking associations even where the specifics of those associations are not yet known and by persuading others that those associations are indeed real (Baake 2003). More specifically, in genetics, plenty of metaphors exist and influence the publics' uptake of genetics (see Chap. 5, this book; van Dijck 1998; Condit 1999; Nerlich et al. 2002; Nelkin and Lindee 2004; Petersen 2005; Giles 2008). Condit (1999), for example, synthesized the 1900s–1990s media reports of genetics into four metaphors, each with respective connotations: the stock-breeding metaphor turned genetics into an enterprise that cultivates "fit" members for the society; the controlling gene metaphor urged families to seek best "odds" for their offspring; the frontier exploration metaphor fanned research experiments and public enthusiasm; and the blueprint metaphor encouraged more holistic, genome-wide examination.

In these studies, however, we have focused almost entirely on verbal metaphors and overlooked visual metaphors. The latter, as I argue below, represents another important venue for analysis. Generally speaking, visual metaphors work in similar ways as verbal metaphors: Two concepts, illustrated through images (the source image and the target image), are presented in such a way "that one idea is used to organize or conceptualize the other" (Kaplan 1992, p. 198). This technique is frequently used by the advertising industry. For example, when an advertisement juxtaposes a shoe with a shoe-sized bed, it suggests that the shoe is like a bed in terms of comfort, warmth, or other essential "bed qualities" (Phillip and McQuarrie 2004).

While the general concept of a visual metaphor is relatively straightforward, there are more specifics to its physical structure and meaning-making mechanism. From a structural perspective, the source image and the target image in a visual metaphor may be separately presented, creating a *juxtaposition metaphor* as in the shoe advertisement mentioned above; alternatively, the source and target images may be combined into one, creating a *fusion metaphor*—an example is fusing a credit card (the target) onto a computer circuit board (the source) to indicate that the card enables customers to make important electronic purchases (Phillips and McQuarrie 2004). From a meaning making perspective, a visual metaphor may be a *similarity metaphor* where the source and target images appear to be like each other—the shoe advertisement is again an example; alternatively, a visual metaphor may be a *connection metaphor* where the source and target images appear to be associated but are not necessarily like each other—the credit card advertisement mentioned above is an example, as the card is not said to be like computers, only associated with computers (Phillip and McQuarrie 2004).

These different categories of visual metaphors inform the following discussion, which is organized based on the two common themes used by images found in this study to metaphorically illustrate DNA. The first theme focuses on DNA's role in life, and the second focuses on its role in medicine.

Life Metaphors

The invocation of life is not uncommon in contemporary mass media's verbal framing of DNA: DNA is habitually referred to as the "secret of life" (Condit 1999; Watson 2004), and related initiatives such as the Human Genome Project are said to offer mankind a "book of life" (Nerlich et al. 2002). The prevalence of these verbal metaphors, as Nelkin and Lindee (2004) critiqued, creates a sense of public euphoria toward genetic research and contributes to a deterministic view of genetics as the commander of life.

Compared with such verbal framing, DNA-life visual metaphors can create even more profound implications. This is because visual metaphors are at once more specific and memorable—because of their use of concrete imageries—and at once more perceptually universal—because of their ability to use varying source images. That is, while DNA invariably takes the shape of the double helix (the target image), the source image can be one of many symbols for life that readers are familiar with: the human body, *part* of the human body, diverse animals and plants, or simply the traces left behind by life forms (e.g., footprints).

Figure 4.10 from a *Scientific American* article, for example, shows a pea plant whose tendrils (those thin, curly stems used by the plant to climb onto objects for support) are morphing into an extended double helix. The shining helix rises up from the background and gradually but dramatically takes center stage. Merging the tendril/plant (the source) and the double helix (the target) into one unified image, Fig. 4.10 makes for a fusion metaphor, which is the structure used by most DNA-life metaphors encountered in this study. In other fused examples, we see the double-helix wrapping around the human body, both the apparently healthy bodies and diseased ones (Wickelgren 1998; Johnson and Andrews 2015), or it running across a human face, blending in with facial features (Chamary and Hurst 2009). These fusion

Fig. 4.10 DNA-life visual metaphor. Because of copyright restrictions, the original image used in Kingsley (2009, p. 53) cannot be reproduced here. The illustration used here resembles the original in content and metaphorical representation (color figure online)

metaphors are structurally more complex than juxtaposition metaphors because readers must be able to disentangle the two fused objects before they can process the image (Phillips and McQuarrie 2004). Their frequent use in popular media, then, suggests an implicit trust that public readers *are* well familiar with the DNA double helix and *can* easily disentangle the embedded metaphorical intent.

Such trust, however, may be ill-placed or at least problematic. From the perspective of metaphorical meaning making, it is ambivalent whether these DNA-life visual metaphors are being used as similarity metaphors or connection metaphors. In other words, it is not clear whether they assert "DNA is (animal/plant/human) life" or "DNA is associated with life." This second reading is certainly reasonable, while technically speaking, the first reading is inaccurate. Though life forms depend on DNA for hereditary information, life as we know it cannot exist without a multitude of other components such as ribosomes (which allow the formation of proteins) or cytoplasm (which is the fluid habitat that houses all cellular parts). But none of these other components can remotely compete with DNA for mass media coverage, so they have no place in public memory. Because of this, and in the face of fusion metaphors such as Fig. 4.10 where concrete organisms physically "turn into" DNA, the complexity of life is readily collapsed into a strand of the double helix. The inaccurate reading prevails: DNA has become life and *is* life.

Surely, even so, readers must recognize that this is merely a figure of speech? Both localized studies and large-scale surveys cast doubt on such assumptions. As Condit (2010) reported, public readers tend to understand DNA and genetics via the general concept of heredity but are not familiar with their finer details (Not coincidentally, the DNA-life metaphor, as discussed below, only supports a general understanding but sheds no light on particulars). In addition, according to the National Science Foundation, as of the early 2000s, only 29% of US adults were able to offer a definition of DNA (National Science Board 2000). More recently, in BBVA Foundation's international studies (2012), only 40% of US adults self-identified as having a complete understanding of DNA. (One should also keep in mind that such data are subject to self-assessment inflation.)

With this backdrop, the "DNA is life" image and message may well be taken literally. What is problematic with this substitution is not just a matter of technical inaccuracy—though that is part of it. What is more problematic is that illustrations like Fig. 4.10 appear to offer fundamental insights without actually doing so. By isolating DNA from its cellular environment and reducing it to two base strands and "rungs" without molecular details (more about molecular details see Chap. 7), they safely ignore the workings of DNA and its interactions with other cellular components. Moreover, by fusing an essentialized double helix with diverse, magnificent life forms, these images appear to offer "a magical revelation" about "life, the universe, and everything'" (Nerlich et al. 2002, p. 465). In Petersen's (2001) words, "the metaphorical and literal have become blurred, so that it is difficult to recognize the ways in which the metaphors invite certain interpretations and not others" (p. 1262).

From a narrow, cognitive standpoint, this "blurring" obscures DNA as a biological entity. Figure 4.10, by fusing the double helix and a physical pea plant with observable leaves, flowers, and peapots, risks conceptual confusions as to what DNA really is: what it is made of, where it resides, or whether it is visible to the unaided eye. Moreover, in trapping, among other things, naked human bodies within the DNA grooves, in amplifying DNA as the one and only entity that matters in life, in suggesting that DNA *is* life, these images eclipse other biological factors and social, cultural, and environmental contexts. The same genetic determinism discussed in Chap. 2 becomes palpable. While in Chap. 2, photographic images encourage this mentality via visceral, emotional appeal, metaphorical illustrations do so, as Lakoff and Johnson (2003) would say, by fundamentally altering the way we conceptualize and react to the world around us. When the double helix is frequently elevated to the same visual plane and magnitude as physical life forms and frequently equated to those life forms, the metaphor is no longer a metaphor or has become a "dead metaphor" whose figurative meaning is shared public knowledge.

What is ironic with this shared knowledge is that it says preciously little other than a superficial and circular catchphrase: DNA is life and life is DNA. All seem to "make sense," but very little actually is. This is the kind of intellectual narcotic and false security that Macdonald (2004) was concerned about in the public communication of science. In this case, readers are offered a grandiose statement about DNA that is easy to illustrate and repeat but carries no operational meaning in a scientific context and only problematic meaning in a social context. As these images continue to affirm shared knowledge, they serve to satisfy curiosity but discourage active inquiries.

Medicine Metaphors

Focused media coverage of biotechnology and genetic medicine emerged in the USA in the late 1970s and early 1980s (van Dijck 1998; Condit 1999). Since then, reports on finding the culprit genes for various diseases and achieving breakthroughs in genetic medicine have become commonplace, a probably inevitable trend given Americans' keen interest in new medical discoveries (see National Science Board 2014). Though frequent, these reports have not, as in the case of DNA-life discussions, resulted in apt verbal metaphors—probably because to say "DNA/genetics is medicine" sounds too simplistically unattractive. The same is not true of visual metaphors: with their ability to portray concrete yet symbolic images, visual metaphors allow readers to engage in an interpretative process about genetic medicine and gain a sense of intellectual satisfaction, without the images, as shown below, actually elucidating the workings and promises of their subject.

Consider Fig. 4.11, which is a National Human Genome Research Institute featured image where the DNA double helix (the target) is fused into a common symbol for medical diagnosis and intervention, the stethoscope (the source). The stethoscope's tubes twist around, merging into a rainbow-colored double helix. While the image's metaphorical intent is obvious, its specific meaning is much more ambivalent. Does it function as a connection metaphor to suggest that genetics and genetic research, symbolized by DNA, are connected with medical diagnosis and intervention? But if so, how are they connected? Or does the image function as a similarity metaphor and suggest that genetics and genetic research *are* the new medicine of today and tomorrow? But, again, if so, in what ways?

Public readers relate to modern medicine via such aspects as visiting doctors, taking medications, and undergoing surgeries—aspects that can be reasonably symbolized via a stethoscope. But readers cannot be assumed to be familiar with the inner workings of pathology, prognosis, and pharmacology—aspects that cannot be reasonably symbolized via a stethoscope and aspects where genetic research, via different processes, actually aids or informs traditional medical techniques. In other words,

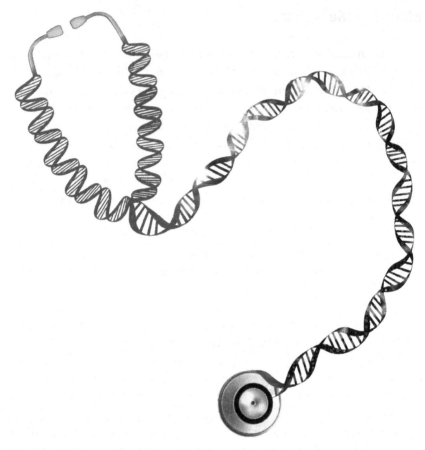

Fig. 4.11   Genetic medicine visual metaphor (Ades 2007) (color figure online)

Fig. 4.11, rather than using a concrete, familiar source image to map onto a target, is using an abstract and complex source to try to shed light on another abstract and complex target. By constructing an apparently metaphorical structure, the image creates the appearance of meaning making—without actually doing so.

Such ambiguous visual metaphors that connect genetics with medical conditions or treatments are common in contemporary popular reports. In Fackelmann (1995), for instance, a strand of the double helix is fused with the rising smoke of a cigarette to, supposedly, illustrate the interplay of genetics and environmental factors in causing cancer. Or, in Saey (2011), an article on the genetic origin of cancer, the double helix is "picked up," by a pair of tweezers held in a human hand, from what one assumes must be cancerous cells. These and other such images, rather than elucidating how genetic research informs modern medicine, actually discourage analytical visualization: what more needs to be said when we can *see* that genetic research is fused with traditional medical equipment, that genetic factors intertwine with carcinogenic environmental factors, or that DNA is physically isolated from diseased cells for targeted examination?

What is especially noteworthy about these visual representations is that they do *not* appear simplistic. Because readers must make an effort to decipher the figurative message embedded in the metaphors, the process can be cognitively satisfying. As Phillip and McQuarrie (2004) wrote, other things being equal, consumers "respond with pleasure to a picture that artfully deviates from expectation. That pleasure arises, in part, from successfully elaborating upon the picture and solving the puzzle it presents" (pp. 126–127). Not unlike visual metaphors used in commercial advertising, the genetic medicine metaphors encourage readers to solve a semiotic puzzle and thus promote positive reactions to the subject depicted, all without quite articulating or explaining the commendable features or qualities of genetic medicine.

If we are concerned with media's one-sided or overly positive framing of genetic research and its influence on public opinions (see van Dijck 1998; Petersen 2001; Nelkin and Lindee 2004), then so should we pay attention to these ambiguous visual metaphors that, in an unassuming but fundamental way, orient readers' attitudes and reactions. Judging by these images, the application of genetic research (never mind what exactly) in modern medicine is natural, imminent (if not already here), and effortless. Brushed aside are those concrete methods, contingent factors, and potential risks that publics, especially minority populations who are excluded from current research (see Chap. 2), are unfamiliar with: For example, details regarding the interactions of genetics and environmental factors, the meaning of genetic risks for disease conditions, the mechanism of genetic testing, and the workings of genetic therapy (see Donovan and Tucker 2000; Kinney et al. 2001; Press et al. 2001; Catz et al. 2005; Kessler et al. 2007; Condit 2010).

As genetic testing and therapy become integrated into clinical practice and the concept of personalized medicine becomes popular, it is such finer-grained knowledge, made relevant to individual consumers, as opposed to a vague awareness of the promise of genetic medicine that will help publics make informed decisions about health care. Otherwise, as Press et al.'s (2001) study showed, publics may harbor rather unrealistic expectations for genetic testing and therapy: In their study, women participants expected that breast cancer testing would have high predictive certainty and could be followed by effective and noninvasive preventive therapy. But in reality, today's genetic research can only offer the opposite: tests of low predicative certainty followed by invasive and limited treatments (Press et al. 2001). As we consider these real-life medical scenarios and the lives of impacted individuals, visual metaphors' ambiguous but effective promotion of genetic medicine becomes more ethically vexing.

By the above, I am not suggesting that visual metaphors are to be avoided in the popular communication of genetics. By using artistic inventions to intrigue readers and concrete images to relate to readers, this visualization technique is valuable for facilitating understanding and engagement. Not to mention that as a rhetorical strategy, metaphors can help simplify complex information and make relevant abstract concepts. But the ideal metaphor, I argue, needs to go beyond clever inventions to consider both cognitive functions and social implications. At the very least, it needs to employ concrete source images whose meanings and intended mapping to the target images are familiar to public readers or are otherwise easily identifiable.

If this sounds like a tall order in the realm of science communication, it is certainly not impossible. *Cell*, a peer-reviewed journal that publishes research on molecular/cell biology, genetics and genomics, and other branches of the life sciences, frequently uses informative as well as creative images, including visual metaphors, for its cover art. A 2013 issue, for example, features two children merrily playing a seesaw game; the children wear traditional Chinese clothes and hair styles, and the playground features the Chinese celebratory color red and traditionally designed kites. The image is unique, bright, and eye-catching; at the same time, it conveys the key message of the issue's leading article, which proposes a seesaw model to pluripotency. Put simply, pluripotency refers to the potential for cells to develop into diverse organs or tissues. Conventionally, it is believed that certain reprogramming factors help to induce pluripotency, while rival factors, working in the opposite direction, induce cells into specific forms (Shu et al. 2013). The seesaw model challenges this view and suggests instead that different rival factors, factors that promote different cell specifications, when in a balanced state, actually promote cell pluripotency (Shu et al. 2013). The two children depicted in the image, then, represent rival factors, and the game is a metaphorical depiction of the seesaw model where balance promotes pluripotency. Given that the research is published by a group of Chinese scientists, the image's stylistic choices also make cultural sense.

Conclusion

Scientific illustrations are not isolated from the contexts of scientific research and the social uptake of that research. In the popular communication of genetics, illustrations grew from minimalist depictions of observed reality and austere teaching of research findings to elaborate and stylized displays as well as figurative inventions. This change reflects the reality of today's mass media industry and acknowledges the need to attract publics, especially large numbers of the publics, to science communication. Everything else being equal, illustrations that appear familiar and inviting can help to increase readers' emotional interest, motivation, and confidence and thus facilitate public understanding and engagement. Even if these illustrations go against conventional information design theories and, in Christiansen's (2013) words, "may distract a bit of attention from communicating the core concept," they seem to have, for all intents and purposes, been accepted as a worthwhile "trade-off" for their affective benefits.

While these benefits are easy—perhaps too easy—to recognize, what is less articulated in current research are contemporary illustrations' multifaceted implications. As argued in this chapter, increased iconic complexity and creative invention, coupled with the increased abstraction of genetic research, may serve to mystify just as they help to elucidate what Christiansen (2013) called "the core concept." Visual metaphors, in particular, may function as a conversation stopper and promote complacency and superficial familiarity rather than true engagement.

A useful way to consider popular science genetics illustrations and perhaps popular science illustrations in general, I suggest, is not to see artistic elaboration and invention as a "trade-off"—which suggests that it, though somewhat problematic, is ultimately necessary—but rather to see it as one, but only one, possible communication tool and goal. Doing so allows us to consider other past and present tools and goals: whether it is minimalist representations that are conducive to amplifying core concepts, comics-style illustrations that are purposefully self-effacing, or creative inventions that emphasize both cognitive and affective functions. Examining, selecting from, and possibly juxtaposing multiple such tools and goals allows scientists and science communicators to more critically consider the contemporary taken-for-granted illustration style and also encourages communication scholars to explore areas for productive research.

Notes

1. This idea is drawn upon Tufte (2006), who cautioned that when arrows do not indicate specific actions or relationships, they only serve to create ambiguities and prevent meaningful analysis.
2. In two-point perspective, a drawing assumes two vanishing points on the horizon line. The top and bottom lines of any vertical plane in the drawing extend to converge on one of these two points. The technique creates the illusion of a three-dimensional object.
3. It should be noted that the literature on this topic is not consistent. Some studies on seductive details yielded conflicting or contradicting results (Schraw 1998; Park and Lim 2007); other scholars (Goetz and Sadoski 1995) questioned the experimental and theoretical foundations of studies that dismiss seductive details.
4. These methods "probe" purified and crystallized molecules with X-rays or radio waves. The resulting X-ray diffraction pattern or radio wave

resonance pattern is then used to deduce the distribution of atoms in a molecule and build a model for its structure (RCSB Protein Data Bank 2016).

5. The invocation of the cosmos is not accidental. As van Dijck (1998) wrote, in contemporary popular media, "geneticists are repeatedly referred to as 'astronauts' of the new science, and genetics as a space adventure. On a symbolic level, it is more heroic to be a small person facing a big challenge than to be a person mastering a phenomenon that is too small for the eye to see" (p. 26). Such syntactic details, then, also serve to promote public appreciation of genetic research.

References

Ades, J. (National Human Genome Research Institute). (2007). *Double Helix with Stethoscope*. Retrieved June 20, 2015, from http://www.genome.gov/dmd/img.cfm?node=Photos/Graphics&id=85334.

Amábile-Cuevas, C. F. (2003). New antibiotics and new resistance. *American Scientist, 91*(2), 138–149.

Amare, N., & Manning, A. (2013). *A unified theory of information design: Visuals, text & ethics*. Amityville, NY: Baywood Publishing.

Baake, K. (2003). *Metaphor and knowledge: The challenges of writing science*. Albany, NY: State University of New York Press.

BBVA Foundation. (2012). BBVA foundation international study on scientific culture: Understanding of science. Retrieved July 21, 2017, from http://w3.grupobbva.com/TLFU/dat/Understandingsciencenotalarga.pdf.

Byrne, M. D. (1993). Using icons to find documents: Simplicity is critical. In B. Arnold, G. Van Der Veer, & T. White (Eds.), *Proceedings of the INTERACT '93 and CHI '93 Conference on Human Factors in Computing Systems* (pp. 446–453). New York: ACM.

Catz, D. S., Green, N. S., Tobin, J. N., Lloyd-Puryear, M., Kyler, P., Umemoto, A., … Wolman, F. (2005). Attitudes about genetics in underserved, culturally diverse populations. *Community Genetics, 8*(3), 161–172. doi:10.1159/000086759.

Chamary, J. V., & Hurst, L. D. (2009, June). The price of silent mutations. *Scientific American*, 46–53.

Christiansen, J. (2013). A defense of artistic license in illustrating scientific concepts for a non-specialist audience. In *Communicating Complexity 2013 Conference Proceedings* (pp. 49–60). Rome: Edizioni Nuova Cultura-Roma.

Condit, C. M. (1999). *The meanings of the gene: Public debates about human heredity.* Madison: University of Wisconsin Press.

Condit, C. M. (2010). Public understandings of genetics and health. *Clinical Genetics, 77*(1), 1–9. doi:10.1111/j.1399-0004.2009.01316.x.

Donovan, K., & Tucker, D. (2000). Knowledge about genetic for breast cancer and perceptions of genetic testing in a sociodemographically diverse sample. *Journal of Behavioral Medicine, 23*(1), 15–36.

Fackelmann, K. (1995). Variations on a theme: Interplay of genes and environment elevates cancer risk. *Science News, 147*(18), 280–281.

Ford, B. J. (1993). *Images of science: A history of scientific illustration.* New York: Oxford University Press.

Galison, P. (1998). Judgment against objectivity. In C. A. Jones, P. Galison, & A. E. Slaton (Eds.), *Picturing science, producing art* (pp. 327–359). New York: Routledge.

Giles, T. D. (2008). *Motives for metaphor in scientific and technical communication.* Amityville, NY: Baywood Publishing.

Goetz, E. T., & Sadoski, M. (1995). Commentary: The perils of seduction: Distracting details or incomprehensible abstractions? *Reading Research Quarterly, 30*(3), 500–511.

Greenwood, V. (2013, April). The fastest DNA sequencer, *Popular Science,* 42–43.

Gross, A. G., & Harmon, J. E. (2013). *Science from sight to insight: How scientists illustrate meaning.* Chicago, IL: University of Chicago Press.

Harp, S. F., & Mayer, R. E. (1997). The role of interest in learning from scientific text and illustrations: On the distinction between emotional interest and cognitive interest. *Journal of Educational Psychology, 89*(1), 92–102.

Harp, S. F., & Mayer, R. E. (1998). How seductive details do their damage: A theory of cognitive interest in science learning. *Journal of Educational Psychology, 90*(3), 414–434.

Image from the RCSB PDB (www.rcsb.org) of PDB ID 2C35. (Meka, H., Werner, F., Cordell, S. C., Onesti, S., & Brick, P. (2005). Crystal structure and RNA binding of the Rpb4/Rpb7 subunits of human RNA Polymerase II. *Nucleic Acids Research, 33,* 6435–6444).

Johnson, R., & Andrews, P. (2015, October). The fat gene. *Scientific American,* 64–69.

June, C., & Levine, B. (2012, March). Blocking HIV's attack. *Scientific American,* 54–59.

Kaplan, S. (1992). A conceptual analysis of form and content in visual metaphors. *Communication, 13*, 197–209.

Katz, M. G., Kripalani, S., & Weiss, B. D. (2006). Use of pictorial aids in medication instructions: A review of the literature. *American Journal of Health-System Pharmacy, 63*(23), 2391–2397.

Kelly, J. B. (1922, March). The physical basis of heredity. *Scientific American*, 177–178.

Kessler, L., Collier, A., & Halbert, C. H. (2007). Knowledge about genetics among African Americans. *Journal of Genetic Counseling, 16*(2), 191–200.

Kingsley, D. M. (2009, January). From atoms to traits. *Scientific American*, 52–59.

Kinney, A. Y., Devellis, B. M., Skrzynia, C., & Millikan, R. (2001). Genetic testing for colorectal carcinoma susceptibility. *Cancer, 91*(1), 57–65.

Lakoff, G., & Johnson, M. (2003). *Metaphors we live by* (2nd ed.). Chicago, IL: University Of Chicago Press.

Lister Hill National Center for Biomedical Communications, U.S. National Library of Medicine, National Institutes of Health, & Department of Health & Human Services. (2016). *Help me understand genetics: Mutations and health*. Retrieved July 17, 2016, from https://ghr.nlm.nih.gov/primer/mutationsanddisorders.pdf.

Macdonald, S. (2004). Authorising science: Public understanding of science in museums. In A. Irwin & B. Wynne (Eds.), *Misunderstanding science? The public reconstruction of science and technology* (pp. 152–171). Cambridge: Cambridge University Press.

Malamed, C. (2009). *Visual language for designers: Principles for creating graphics that people understand*. Beverly, MA: Rockport Publishers.

McCloud, S. (1994). *Understanding comics: The invisible art*. New York: HarperCollins.

McDougall, S. J., de Bruijn, O., & Curry, M. B. (2000). Exploring the effects of icon characteristics on user performance: The role of icon concreteness, complexity, and distinctiveness. *Journal of Experimental Psychology: Applied, 6*(4), 291–306.

Miller, G. A. (1994). The magical number seven, plus or minus two: Some limits on our capacity for processing information. *Psychological Review, 101*(2), 343–352.

Misteli, T. (2011, February). The inner life of the genome. *Scientific American*, 66–73.

Molyneaux, B. L. (2013). Introduction: The cultural life of images. In B. L. Molyneaux (Ed.), *The cultural life of images: Visual representation in archaeology* (pp. 1–10). London: Routledge.

National Institutes of Health. (2009). New NIH research plan on fragile x syndrome and associated disorders. Retrieved June 20, 2015, from http://www.nichd.nih.gov/news/resources/spotlight/Pages/071609-Fragile-X.aspx.

National Science Board. (2000). *Science and engineering indicators 2000.* Arlington, VA: National Science Foundation.

National Science Board. (2014). *Science and engineering indicators 2014.* (No. NSB 14-01). Arlington, VA: National Science Foundation.

Nelkin, D., & Lindee, M. S. (2004). *The DNA mystique: The gene as a cultural icon.* Ann Arbor: University of Michigan Press.

Nerlich, B., Dingwall, R., & Clarke, D. D. (2002). The book of life: How the completion of the human genome project was revealed to the public. *Health: An Interdisciplinary Journal for the Social Study of Health, Illness and Medicine, 6*(4), 445–469.

Nettelbeck, D. M., & Curiel, D. T. (2003). Tumor-busting viruses. *Scientific American, 289*(4), 68–75.

Park, S., & Lim, J. (2007). Promoting positive emotion in multimedia learning using visual illustrations. *Journal of Educational Multimedia and Hypermedia, 16*(2), 141–162.

Peirce, C. S. (1894). *What is a sign?* Retrieved June 18, 2015, from http://www.iupui.edu/~arisbe/menu/library/bycsp/bycsp.HTM.

Petersen, A. (2001). Biofantasies: Genetics and medicine in the print news media. *Social Science & Medicine, 52*(8), 1255–1268.

Petersen, A. (2005). The metaphors of risk: Biotechnology in the news. *Health, Risk & Society, 7*(3), 203–208.

Phillips, B. J., & McQuarrie, E. F. (2004). Beyond visual metaphor: A new typology of visual rhetoric in advertising. *Marketing Theory, 4*(1), 113–136. doi:10.1177/1470593104044089.

Press, N. A., Yasui, Y., Reynolds, S., Durfy, S. J., & Burke, W. (2001). Women's interest in genetic testing for breast cancer susceptibility may be based on unrealistic expectations. *American Journal of Medical Genetics, 99*(2), 99–110. doi:10.1002/1096-8628(2000)9999:999<00:AID-AJMG1142>3.0.CO;2-I.

RCSB Protein Data Bank. (2016). *Methods for determining atomic structures.* Retrieved January 6, 2016, from http://pdb101.rcsb.org/learn/guide-to-understanding-pdb-data/methods-for-determining-structure.

Saey, T. H. (2008). Epic genetics. *Science News, 173*(17), 14–19.

Saey, T. H. (2011). Tumor tell-all. *Science News, 180*(7), 18–21.

Schraw, G. (1998). Processing and recall differences among seductive details. *Journal of Educational Psychology, 90*(1), 3–12.

Seeman, N. (2004). Nanotechnology and the double helix. *Scientific American,* 64–75.

Shirk, H. N., & Smith, H. T. (1994). Some issues influencing computer icon design. *Technical Communication, 41*(4), 680–689.

Shu, J., Wu, C., Wu, Y., Li, Z., Shao, S., Zhao, W., ... Deng, H. (2013). Induction of pluripotency in mouse somatic cells with lineage specifiers. *Cell, 153*(5), 963–975.

Stein, G. S., Stein, J. S., & Kleinsmith, L. J. (1975). Chromosomal proteins and gene regulation. *Scientific American, 232*(2), 46–57.

Sweller, J. (1994). Cognitive load theory, learning difficulty, and instructional design. *Learning and Instruction, 4*(4), 295–312.

Thone, F. (1927). X-rays speed up evolution over 1,000 per cent. *The Science News-Letter, 12*(340), 243–246.

Tufte, E. R. (2001). *The visual display of quantitative information.* Cheshire, CT: Graphics Press.

Tufte, E. R. (2006). *Beautiful evidence.* Cheshire, CT: Graphics Press.

van Dijck, J. (1998). *Imagenation: Popular images of genetics.* New York: New York University Press.

Wagner, E. K. (1974). The replication of herpes viruses. *American Scientist, 62*(5), 584–593.

Ware, C. (2012). *Information visualization: Perception for design* (3rd ed.). Burlington, MA: Morgan Kaufmann.

Watson, J. (2004). *DNA: The secret of life.* New York: Alfred A. Knopf.

Wickelgren, I. (1998, November). Gene readers. *Popular Science,* 56–61.

Yu, H. (2015). *The other kind of funnies: Comics in technical communication.* Amityville, NY: Baywood Publishing.

Yuan, R., & Hamilton, D. L. (1982). Restriction and modification of DNA by a complex protein. *American Scientist, 70*(1), 61–69.

5

The Code View: Cracking the Genetic Code of Life

In popular communication of genetics, it is common to hear that DNA is a "code," that genetic materials speak a mysterious language with "encoded" meanings, and that genetic research is a heroic effort to crack that code (or a doomed attempt to mess with it). The high-profile Human Genome Project, for example, describes its accomplishment as producing a "finished version of the human genetic code" (National Human Genome Research Institute 2010). Francis Collins (2011), current director of the National Institutes of Health, implied this code by titling his popular book on DNA and personal medicine *The Language of Life*. And prominent television program provider the Public Broadcasting Service (2001) equates the "common genetic code" shared between humans and other species to the universal thread of life.

Although various scholars have critiqued the use of the code metaphor as being essentializing and misleading (van Dijck 1998; Nelkin and Lindee 2004), as this study shows, the metaphor is here to stay and occupies a central place in the public discourse about genetics. Accepting this reality, this chapter examines how visual representations have (or have not) been used to depict and unpack this metaphor to public readers and how these representations change in the face of shifting research environments and social contexts. In doing so, the chapter

© The Author(s) 2017
H. Yu, *Communicating Genetics*,
DOI 10.1057/978-1-137-58779-4_5

contemplates the ways these images solidify a grand metaphor for genetics and how they contribute to the social uptake of the discipline.

Unveiling the Original Genetic Code

To be precise, the genetic code metaphor, as Knudsen (2005) pointed out, is not a single metaphor but a network of related expressions created around an initial, root metaphor. While the development of this entire network is difficult to trace, the initial root metaphor is believed to be coined not by a geneticist or biologist but physicist and Nobel Prize winner Erwin Schrödinger (1887–1961). Schrödinger invented the metaphor largely because of his inability to reconcile the findings of genetics and the laws of physics. As he explained in a 1943 lecture at the Trinity College and later in his popular book *What is Life?*, for genes to pass hereditary traits through numerous duplications over thousands of years, they must behave in a rather stable manner (Cobb 2015). This assumption, however, would violate the laws of physics, according to which individual atoms perform in a "disorderly heat motion"; everyday objects, say a table, maintain their physical composition only because they contain enormous numbers of atoms, so the disorder of individual atoms, statistically speaking, leaves no impact on the objects as a whole (Schrödinger 1944, p. 8). But because a gene contains a relatively small number of atoms, it *should not* remain as constant as it needs to be in order to pass on hereditary information over generations (Cobb 2015). To explain this violation, Schrödinger (1944) speculated that an organism must "contain in some kind of code-script the entire pattern of the individual's future development and of its functioning in the mature state" (p. 15).

This code-script metaphor, however, did not become popular until two decades later. van Dijck (1998) suggested that this delayed acceptance was correlated with the 1960s' development of Noam Chomsky's generative linguistics, which sees language as a code system that follows set rules to generate meaning. Also theorized to have contributed to the metaphor's delayed popularity is the World War II and Cold War emphasis on code breaking and cryptanalysis (van Dijck 1998; Knudsen 2005), the emergence of computers, and the integration of computer and information science into molecular biology (Knudsen 2005). But a brief examination of the disciplinary history of genetics would show that the uptake of this metaphor is equally, if not more, related to the breakthroughs in DNA research leading up to the 1960s.

As early as 1919, biochemist Phoebus Levene had proposed that DNA is composed of a series of linked nucleotides; each nucleotide contains a sugar molecule, a phosphate group, and one of four nitrogen-containing bases (Pray 2008). The four bases are adenine (A), thymine (T), guanine (G), and cytosine (C). Extending Levene's work, Erwin Chargaff discerned, in 1950, that the four bases maintain a constant ratio in DNA: the total amount of A is similar to that of T, and the total amount of C is similar to that of G (Pray 2008). Armed with Chargaff's findings and other researchers' work (see Chap. 2), in 1953, James Watson and Francis Crick deduced that a DNA molecule is two strands of sugar-phosphate backbones bound together by the four bases; base A on one strand always pairs with T on the other, and base C on one strand always pairs with G on the other, giving rise to the same number of A and T bases and C and G bases.

In March 1953, two weeks after deriving the DNA model, Crick wrote the following in a letter to his 12-year-old son: "Now we believe that the D.N.A. is a code. That is, the order of the bases (the letters) makes one gene different from another gene" (Cobb 2015, p. 110). Two months later, in a *Nature* article Crick published with Watson, they publicly alluded to this code, stating that it "seems likely that the precise sequence of the bases is the code which carries the genetical information" (Watson and Crick 1953, p. 965). Following this announcement, scientists worldwide joined in the effort to crack this code; notable among them were Marshall Nirenberg, Har Gobind Khorana, and Robert Holley, all of whom received the 1968 Physiology or Medicine Nobel Prize for their contribution. Together and by the mid-1960s, these and other researchers had revealed the mechanism of how DNA base sequences impact protein translation (see Fredholm 2004), effectively cracking the genetic code as it was originally conceived in the last century.

The code turned out to be surprisingly simple—and conducive to creating a network of related metaphorical expressions. Put simply, every three adjacent DNA bases (called a codon) act as *letters* to *spell out* one amino acid; these amino acids, as *words*, are *strung* together to make a protein, or a *sentence*; and the protein-sentences, dispersed in living organisms, give rise to the *book* of life. There are, altogether, 20 common amino acid words in the book. A TTT codon, for example, spells out amino acid Phe, while a TTA codon spells out Leu; several codons are also responsible for *signaling* the start and stop of protein translation. As all *languages*, the code of life contains certain redundancies; that is, multiple codons could mean the same amino acid words or signals.

One of the first ways this code was depicted for public readers was via the codon table (see, e.g., Jukes 1963; Crick 1966), which lists possible codon combinations and their corresponding amino acids/signals. These tables, as in the case of Table 5.1, are frequently expressed from the standpoint of messenger RNA, the "middle person" that transcribes the code out of DNA in preparation for protein translation. Messenger RNA shares the same three bases with DNA—A, G, and C—but its fourth base, instead of a thymine (T), is a structurally similar uracil (U).

Table 5.1 The genetic codon table

UUU → *Phe*	UCU → *Ser*	UAU → *Tyr*	UGU → *Cys*
UUC → *Phe*	UCC → *Ser*	UAC → *Tyr*	UGC → *Cys*
UUA → *Leu*	UCA → *Ser*	UAA → *Stop*	UGA → *Stop*
UUG → *Leu*	UCG → *Ser*	UAG → *Stop*	UGG → *Trp*
CUU → *Leu*	CCU → *Pro*	CAU → *His*	CGU → *Arg*
CUC → *Leu*	CCC → *Pro*	CAC → *His*	CGC → *Arg*
CUA → *Leu*	CCA → *Pro*	CAA → *Gln*	CGA → *Arg*
CUG → *Leu*	CCG → *Pro*	CAG → *Gln*	CGG → *Arg*
AUU → *Ile*	ACU → *Thr*	AAU → *Asn*	AGU → *Ser*
AUC → *Ile*	ACC → *Thr*	AAC → *Asn*	AGC → *Ser*
AUA → *Ile*	ACA → *Thr*	AAA → *Lys*	AGA → *Arg*
AUG → *Met*	ACG → *Thr*	AAG → *Lys*	AGG → *Arg*
GUU → *Val*	GCU → *Ala*	GAU → *Asp*	GGU → *Gly*
GUC → *Val*	GCC → *Ala*	GAC → *Asp*	GGC → *Gly*
GUA → *Val*	GCA → *Ala*	GAA → *Glu*	GGA → *Gly*
GUG → *Val*	GCG → *Ala*	GAG → *Glu*	GGG → *Gly*

Thus, from the standpoint of RNA, a T in the code would be changed to a U. In the early days, some of the codon combinations and functions were still unclear; those were duly noted in the tables.

A codon table, much like a *grammar* book, lays out in plain view the rules in the genetic code of life. It, however, does not quite illustrate the significance of the code as that which maintains hereditary traits from generation to generation or creates diverse (or mutated) life forms and functions. That significance is more easily fulfilled by another form of early code views in popular communication: illustrations of hypothetical DNA bases and corresponding amino acids, as shown in Fig. 5.1. The top half of the figure compares the bases of a "normal" DNA segment with those of several mutated segments. Red frames indicate where mutations are to occur in the normal DNA, and red texts highlight the results of those mutations. The bottom half of the figure then compares the amino acids spelled out by the normal and mutated DNA; amino acids are strung together to form hypothetical proteins. Again, frames and color are used to highlight sites of mutation.

Compared with codon tables, images like Fig. 5.1, by demonstrating the dynamic correlation between DNA and protein, offer more concrete examples of the genetic code at work and its significance.

Normal DNA	A T T	G G G	G A T	T A C	G G C
Mutant 1 DNA	A C T	G G G	G A T	T A C	G G C
Mutant 2 DNA	A T T	G G G	G T T	T A C	G G C
Mutant 3 DNA	A T T	G G G	G A T	T A C	G A C

Normal protein	*Ile*	*Gly*	*Asp*	*Tyr*	*Gly*
Mutant 1 protein	*Thr*	*Gly*	*Asp*	*Tyr*	*Gly*
Mutant 2 protein	*Ile*	*Gly*	*Val*	*Tyr*	*Gly*
Mutant 3 protein	*Ile*	*Gly*	*Asp*	*Tyr*	*Asp*

Fig. 5.1 Base mutation and amino acid change. Created based on Yanofsky (1967, p. 87) (color figure online)

In this particular case, it is notable how a single mutated base can change protein translation and how diverse those changes can be. More importantly, by exemplifying the way the genetic code works, these images help to demetaphorize and thus demystify this grand metaphor. Born out of unresolved dilemmas, the code metaphor carries with it a distinctive air of mystery. For readers who have intimate knowledge of how the "code" works, this single word conveys a network of complex information. But for readers who do not, frequent references to the "code" without corresponding explanation only serve to mystify genetics and genetic research. Images like Fig. 5.1, then, are notable for allowing readers to witness the code for what it is: a biochemical mechanism wherein particular nucleotide acid sequences are transcribed and translated into specified proteins.

In addition to "stand-alone" representations like Table 5.1 and Fig. 5.1, the genetic code was integrated into procedural and schematic illustrations (more about illustrations see Chap. 4). Figure 5.2, for example, demonstrates how "the four-letter 'language' of nucleic acids" is turned into the "language of proteins" (Spiegelman 1964, p. 49). In this example, a string of messenger RNA (which, again, is the middle

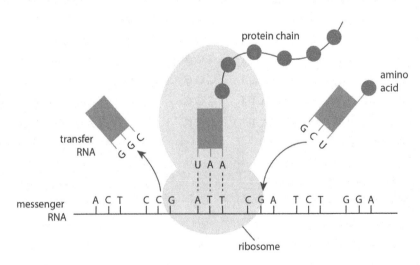

Fig. 5.2 Genetic code integrated into illustration. Created based on Spiegelman (1964, p. 49) (color figure online)

person transcribed from DNA) is being translated, one codon at a time, into protein. The process is enabled by the ribosome, which "holds" the messenger RNA in place and assembles amino acids into a protein chain, and enabled by transfer RNA, which moves in and out to carry specified amino acids to the site.

Images like Fig. 5.2, by integrating the abstract genetic code into iconic and thus more concrete illustrations of genetic activities and processes, help to communicate the context of the code at work. Together with codon tables and displays of hypothetical bases vis-à-vis amino acids, they offer a rich and cognitively interesting picture of the genetic code as it was conceived and cracked back in the 1960s.

Traditional DNA Sequencing: Peeking at the Expanded Code

With the DNA-protein-coding mechanism resolved, the next step was to determine individual organisms' actual DNA base sequences so as to pinpoint (and influence) the "messages" written into those sequences. In many ways, these efforts foreshadowed today's expanded genetic code metaphor in which we went beyond considering DNA as a language with four base letters and 20 amino acid words to envisioning an organism's entire, enormous DNA (its genome) as a complex system of biological information. Providing backdrops for this new vision is, of course, the emergence of DNA-sequencing technologies.

The earliest success in DNA sequencing came about in the 1970s. Notably, in 1973, Walter Gilbert and Allan Maxam sequenced a short, 24-base-long segment of the *Escherichia coli* bacterial DNA (Gilbert and Maxam 1973). The method they used, known as wandering-spot analysis, works by trimming away a DNA segment one nucleotide at a time with an enzyme and thereby obtaining a series of breakdown products of all possible lengths. These products are separated by size to appear as "stained" spots. The sequence of the original segment is then deduced based on the spots' dispersed, or "wandering," pattern. This method, however, is time-consuming and labor-intensive (Adams 2008). In 1977, Gilbert and Maxam (1977) came up with a new technique, which uses different chemical agents to cleave a DNA segment at the four bases; i.e., one agent would cleave the segment at all the A bases, another at all the T bases, and so on. By comparing the four sets of resultant DNA fragments via electrophoresis (see Chap. 2), one can deduce the position and identity of each base.

In the same year, Frederick Sanger and his colleagues reported a more efficient and accurate sequencing method known as chain termination or, simply, the Sanger method (Sanger et al. 1977). This method attempts to synthesize a DNA segment based on the template of an original one and then uses targeted inhibitors to terminate the synthesis at each base, i.e., at all the As, all the Ts, and so on. By comparing the resultant fragments via electrophoresis, one can deduce the original DNA sequence. The Sanger method, known as traditional or first-generation sequencing, was the gold standard for DNA sequencing for the next 25 years (Grada and Weinbrecht 2013). It also paved the way for automated sequencing that uses fluorescent labels, laser-induced signal detection, and computerized analysis to more efficiently identify DNA bases. These automated machines started to appear in the mid-1980s, enabling still cheaper and faster processes (Adams 2008). Sanger, together with Gilbert, was awarded the 1980 Nobel Prize in Chemistry for their work.

With these developments, various organisms' actual DNA sequences started to appear in popular communication. Figure 5.3, for example, shows 1567 DNA bases from the mouse genome. This segment is associated with the mouse hemoglobin protein, which is responsible for transporting oxygen through blood. Not all bases shown in the figure, however, actually code for amino acids. As the original caption explained, from the base labeled cAp on the first line to the base labeled pA on the next-to-last line, all the bases in between are transcribed into the "middle person" messenger RNA. However, those non-highlighted bases (known as noncoding bases) are then cleaved, leaving only the highlighted ones to code amino acids and form the protein. Respective amino acids are written below their bases.

As Fig. 5.3 shows, displaying part of an organism's DNA, let alone its entire genome, takes considerable physical space. In the 1970s–1980s popular science magazines, entire page or pages were at times devoted

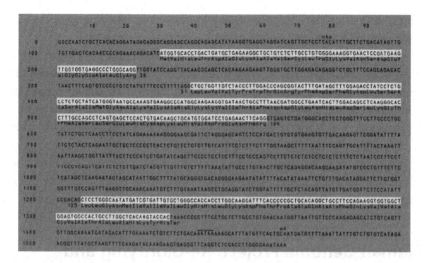

Fig. 5.3 Extended base sequence from the mouse genome (Wetzel 1980, p. 669). Adapted and reprinted from *Cell*, 15/4, Konkel, David A.; Tilghman, Shirley M.; Leder, Philip, The sequence of the chromosomal mouse β-globin major gene: Homologies in capping, splicing, and poly(A) sites, 1125-1132, Copyright (1978), with permission from Elsevier

to this purpose. When the complete DNA sequence of the φX 174 virus was determined, for example, the 5386 bases spread across the good part of six pages of an *American Scientist* article (Smith 1979), which is unthinkable today when popular reports feature, at most, double-spread images. Images that warrant this "real estate" must also, by unspoken rules, be visually exciting, which the rows upon rows of DNA bases hardly qualify for.

Of course, extensive sequence images like Fig. 5.3 were products of their own time and disciplinary context. As DNA-sequencing technologies were a breakthrough at the time, these images were not merely displays of certain localized experimental results but evidence of methodological success that contributed to the larger genetic research agenda. The same sentiment and acknowledgement were evident in professional journal publications at the time. When Gilbert and Maxam (1973) first sequenced the 24 bases of the *E. coli* genome, those base pairs were prominently displayed in the article's short abstract. When longer sequence data became available, they were likewise provided in their entirety in journal publications (see, e.g., Coleman et al. 1987).

Although images like Fig. 5.3 may lack visual appeal, they are arguably (and comparatively, as we will see later) concrete visual representations of the now expanded genetic code metaphor. By depicting actual compositions of DNA and proteins, they provide readers with a way to imagine how genetic information is "encoded" in organisms. This is especially the case when these images also explain, in context, finer-grained "coding" features. In Fig. 5.3, for example, one discerns the difference between protein-coding and noncoding bases and how these bases may be scattered inside a genome.

Human Genome Project: Re-Glorifying and Re-Mystifying the Code

With the advancement in sequencing technologies, in the 1980s, researchers started to consider the possibility of sequencing the entire human genome so as to obtain a complete human DNA "blueprint." This "blueprint," it was hoped, would then reveal the secret of human life and afford us the ability to maintain and regulate that life. However, despite the refined techniques at the time, sequencing remained a time-consuming and expensive task. Sequencing the enormous human genome was beyond the capability of any individual researchers or laboratories. That is where the National Institutes of Health and Department of Energy came in. In 1988, the two agencies received Congress funding to explore the human genome, and in 1990, they published a joint research plan titled "Understanding Our Genetic Inheritance: The Human Genome Project," which set goals for the first 5 years of what was, at the time, a 15-year research plan (National Human Genome Research Institute 2012). Various other organizations, notably The Wellcome Trust based in the UK and private company Celera, were also involved in the Human Genome Project (Dovichi and Zhang 2000).

Though the project is most well known in popular media as an effort to sequence the human DNA, it is also a project intended to develop better, cheaper, and faster sequencing methods. The project led to—and was enabled by—various developments in sequencing technology: for example, more efficient and sensitive fluorescent dyes that can produce more accurate results (Metzker 2005), and capillary electrophoresis that uses tiny silica tubes to achieve fast and simultaneous processing of large numbers of DNA samples (Dovichi and Zhang 2000; Mitchelson et al. 2011). With these developments, the Human Genome Project was completed ahead of schedule in 2003, after having sequenced the 3 billion base pairs in the human genome at a cost of $2.7 billion dollars. In addition to the human genome, the project also sequenced a number of genomes important for animal model studies, including the mouse, rat, and fruit fly genomes (National Human Genome Research Institute 2011).

Post Human Genome Project, the code metaphor has irrevocably expanded. The codon-amino acid translation mechanism faded into the background, while "code" is elevated to be the entirety of genomic data and the whole of biological information embedded in those data. It is now commonplace for media reports to conflate sequencing and decoding. The sequencing of the wheat genome, for example, is hailed in *USA Today* as a success in cracking wheat's genetic code (Satter 2010), and personal genome sequencing is compared to obtaining one's own version of the code (Harmon 2008). Even in professional publications, researchers use metaphors such as the "second genetic code" or "decoding the non-coding" (see Tejedor and Valcárcel 2010; Schonrock and Götz 2012) to situate their work beyond the 1960s code. This expansion brings more and renewed glamor to the grand code metaphor as media worldwide equate this concept to life's ultimate secret and celebrate its value in scientific research and medical application (also see Petersen 2001; Nerlich et al. 2002).

But what is ironic is that this expansion has served to re-mystify the code enigma that earlier works helped to peel away. That is, the findings of the Human Genome Project and subsequent studies led researchers to conclude that despite obtaining whole-genome sequences, we know very little about what those sequences actually mean. For example, the majority of the human DNA (about 97% of it) does not actually code amino acids; in other words, it does not follow the original code that was cracked back in the 1960s. The same is true, to varying degrees, with other organisms. These large portions of DNA, known as noncoding DNA (see Fig. 5.3), were once thought to be useless "junk" but are increasingly recognized to encode biochemical functions; importantly, they are transcribed into noncoding RNA that in turn regulates various genetic activities (Mattick and Makunin 2006; Dunham et al. 2012). The exact functions and mechanisms of these vast amounts of noncoding DNA and RNA, however, remain unknown and contested.

All of these myths—and glories—about the genetic code found their way into popular visual representations, which are a far cry from earlier images that try to (to the extent possible) materialize the code into concrete bases and amino acids. Saying little about what we do know about the code and romanticizing what we do not know, these images seem more interested in channeling and fostering viewer reactions than providing actual information. In that respect, they resemble the symbolic photographs (Chap. 2) and illustrations (Chap. 3). Figure 5.4 is a typical example. In this case, silhouettes of a group of people of different statures and presumably different genders are composed, in a collage

Fig. 5.4 Silhouettes of people comprised of DNA base letters (Ades 2005) (color figure online)

style, of strings of ATCGs. These DNA bases appear in bright and seemingly random colors against a darker background; it is unknown whether the bases represent actual human genome sequence snippets.

Similar, and often more elaborate, portrayals of the code are common in popular communication where sequence letters are used to compose anything from human faces, plants, to animals and in styles ranging from water color to cartoons (see Sinha 2000; Cevallos 2011; Rosen 2015). In one of the more striking examples, a full *Scientific American* magazine page is covered by RNA bases arranged in three-letter codons (Freeland and Hurst 2004, p. 85). The codons appear in different shades of green and contribute to the optical illusion of a human figure. The "green code" is distinctly reminiscent of the "green screen," the contemporary symbol for mysterious, omnipresent, postmodern high-tech—think of *The Matrix* (Sosinsky 2008, p. 262).

Even more notable are high-profile examples at "Genome: Unlocking Life's Code," an educational exhibit jointly produced by the National Human Genome Research Institute and the National Museum of Natural History. Here, the human DNA sequence is projected as "an endless scroll…on a huge TV screen" and worn, in "eye-popping orange," by a female mannequin aptly nicknamed Gigi (Skirble 2013). According to the exhibit director Meg Rivers, Gigi (Fig. 5.5) "has on her the human genetic code and she is meant to really engage visitors and get them to think about having their own code and what does it mean to them [sic]" (Skirble 2013).

By creatively and artistically combining life forms (mostly humans) with genetic sequences in these ways, contemporary code images appear far more exciting and attractive than those from earlier decades. By physically fusing life forms with the code, they prompt readers to acknowledge the equivalence of the two. And by invoking splendid, mysterious high-tech illusions, they set genetics apart from other (mundane) sciences as a cool and awe-inspiring endeavor. Conspicuously missing from these images, however, are attempts to explain, unpack, or contextualize the genetic code. Compared with earlier representations, these images do precious little (whether within the images themselves or in accompanying captions) to explain what the endless "letters" stand for, how they relate to genes and other gene products, how they

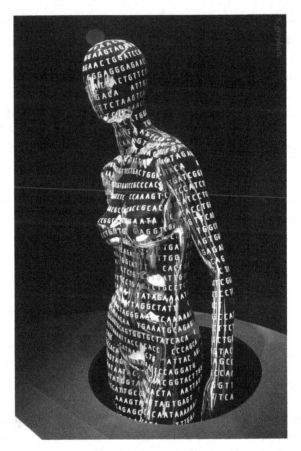

Fig. 5.5 Gigi wears the human genetic code (Human Genome mannequin 2013). Mannequin created by Evidence Design (color figure online)

"encode" human and other life forms, and what their significance is. Given these, it seems unlikely that seeing them would allow readers to "think about having their own code and what does it mean to them" (Skirble 2013).

Indeed, it is arguable that these contemporary representations serve to mystify the genetic code as that which is beyond public grasp and thus impossible or unnecessary to explain in popular media. It suffices, so it seems, to simply combine molecular-level genetic components with physical life forms and call that a good attempt. As van Dijck (1998) also pointed out, although "code" has become a conceptual framework for thinking about genetics as information processing and should allow notions of complexity, "the images that evolve along with information processing systems reflect anything but complexity" (p. 123). She cited Walter Gilbert, chief proponent of the Human Genome Project, making this enthusiastic remark: "Three billion bases of sequence can be put on a single compact disk, and one will be able to pull a CD out of one's pocket and say 'Here is a human being, it's me'" (van Dijck 1998, p. 123). The CD, as the human bodies and mannequin composed of sequence data, collapses tremendous technical details and assumptions, leaving readers with no entry point to broach the code.

In addition, it is easy to see how these contemporary representations, by literally reducing human bodies and other life forms to DNA/RNA bases, fortified genetic essentialism and neurogenetic determinism (more about these sentiments see Chap. 2). To borrow Nelkin and Lindee's words (2004), in these images, genetic sequence and code become "an invisible but material entity," "a sacred entity, a way to explore fundamental questions about human life, to define the essence of human existence, and to imagine immortality" (p. 40). Physical life forms no longer matter; what matters is the continuation of the code.

Certainly, these images are not the only way contemporary mass media depict the genetic code. Images that build upon and expand earlier works also exist. In Drayna (2005), for instance, we see rows of hypothetical DNA sequences compared to demonstrate the difference between a founder mutation (which happens in the ancestor of a population and is inherited by many individuals) and a hot spot mutation (which arises spontaneously in different individuals). Such depictions, similar to Fig. 5.1 shown earlier, emphasize DNA base data without a celebratory or mysterious undertone. And as with earlier works, contemporary images also integrate the genetic code, in its expanded sense, into procedural and schematic illustrations. Figure 5.6, for example, explains polymerase chain reaction, a common procedure that duplicates a target DNA strand into vast copies to enable subsequent analysis. During the procedure, a short DNA segment called a primer pairs up with a target strand; an enzyme called DNA polymerase then works from the primer to make a complete

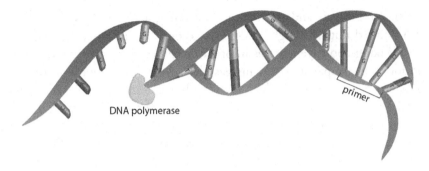

DNA polymerase

primer

Fig. 5.6 Genetic code integrated into illustration: Contemporary take. Created based on Adleman (1998, p. 58) (color figure online)

complement of the target strand. The two strands will then separate and be used as templates to create additional copies, and so on and so forth, resulting in a "magnification" of the original code.

Figure 5.6, as earlier such works (see Fig. 5.2), affords a more contextualized representation of the genetic code at work. In addition, it is more syntactically complex (with colors and dimensions) and more visually attractive. These syntactic details, however, also make it difficult for one to read, let alone examine or compare, the sequence data amidst colorful, spiraling, and dimensional ribbons and base columns. Moreover, the ubiquitous use of such details makes it more likely for images to reduce the actual sequence data they integrate. In one *Popular Science* article (Skloot 2004), for instance, a single base letter within a sequence of four (superimposed on colorful, three-dimensional columns) is used to represent the genetic marker that can suggest an individual's ancestral geographic origin and ethnicity—when realistically, a set of a hundred or so markers may be needed for such purposes (Nassir et al. 2009).

Although visual choices alone do not speak to the paradigm of popular science communication, in the case of the genetic code and compared with earlier works, contemporary representations, whether the creative collages or colorful code-illustration hybrids, exhibit a general tendency to at once embellish and simplify. It is this tendency that results in attractive but uncomplex visual representations. Of course, this tendency occurs only in popular communication. In specialist venues, as seen below, advancements in sequencing technologies have ushered in, and continue to afford, more complex, efficient, and versatile visualizations.

Second-Generation Sequencing: Specialist Visualization

Following the completion of the Human Genome Project, various new sequencing technologies emerged that depart from the first-generation Sanger method, including, for example, pyrosequencing and sequencing by synthesis using reversible terminators. Often referred to as next- or second-generation sequencing, these technologies generally use massively parallel processes to dramatically increase throughput, raise speed, and lower cost (Mitchelson et al. 2007; Grada and Weinbrecht 2013). Grada and Weinbrecht (2013) estimated that it costs around $100,000 to set up a next-generation sequencing platform and then upward of $ 1000 to sequence a small genome within the space of one day, which are astronomical improvements compared to the decades and billions of dollars invested in the Human Genome Project. It is hoped that next-generation sequencing can provide researchers and clinicians with a practical means to sequence patients' protein-coding DNA in order to identify disease-causing mutations or perform targeted studies on disease-causing genomic regions (Grada and Weinbrecht 2013).

While this vision of a DNA-centered future of biological studies and health care has yet to fully realize, next-generation sequencing certainly prepared for it by generating an enormous amount of sequence data. To present, analyze, and share these data became a pressing need and led to the development of digital databases and data visualization platforms around the world, with the most notable ones housed by the National Center for Biotechnology Information (NCBI). A division of the National Library of Medicine, NCBI provides a range of databases and tools that enable users to search for population-specific, organism-specific, gene-specific, or mutational sequences. With a given sequence, users can then choose to compare it with another sequence, to search for subsequences or patterns, or to highlight targeted features.

160 H. Yu

Figure 5.7, for example, compares a gene called cytochrome oxidase subunit I (COI) between two catfish species (*Bagarius suchus* and *Bagarius bagarius*); this gene encodes a protein that is involved in the fishes' energy metabolism. Because the COI sequence is over 600 DNA bases long and cannot all be displayed at once, the top panel of Fig. 5.7 provides a navigation bar. It allows users to see which region of the sequence they are viewing and to select different regions by dragging a frame along the bar. Below the navigation bar, the sequence for *B. Suchus'* COI gene is displayed and shown in two complementary strands. The amino acids coded by the top strand, together with that same strand organized in three-letter codons, follow below. Last, in the bottom panel, *B. bagarius'* COI sequence is given for cross-species comparison, with their occasional differences highlighted in red. Beyond these features shown, users are allowed to make custom configurations such as removing or adding certain panels, setting a marker at a particular sequence location, or changing the size/color of base letters.

The advantage of such digital representations is obvious when they are compared with earlier, print-based images that, though commendable in their effort to provide concrete sequence reads, are cumbersome to view and search. If Figs. 5.1 and 5.3 shown earlier are not fully illustrative of these issues, consider Fig. 5.8. This 1960s code image illustrates different kinds of DNA mutation and how the resultant mutants may subsequently recombine. In part (a) of the image, a normal,

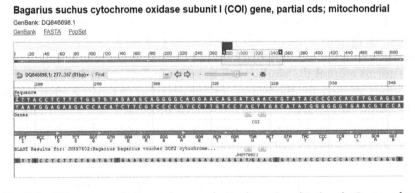

Fig. 5.7 COI gene of *B. suchus* and *B. bagarius* (National Center for Biotechnology Information 2016b) (color figure online)

double-strand DNA segment undergoes, separately, point mutation, deletion mutation 1, and deletion mutation 2. In part (b), the resultant strands of point mutation and deletion mutilation 1 recombine to form a normal sequence. In another part not shown here, the resultant strands of point mutation and deletion mutation 2 recombine but does not form a normal sequence.

Even though Fig. 5.8 uses only a short DNA segment, to tell all of its stories still resulted in a complex mix of letters, colors, frames, lines, and arrows that discourages data comparison. In part (a), while it is relatively easy to compare the normal DNA with the adjacent point mutation, it is less clear what exactly happened in the two deletion mutations several lines below. In part (b), it gets more confusing as to how one should interpret the different lines and arrows. Such presentation challenges are much easily handled by digital tools with their ability to zoom in for detail and zoom out for context and to offer customized search, comparison, and display. These merits are what motivate the continuous

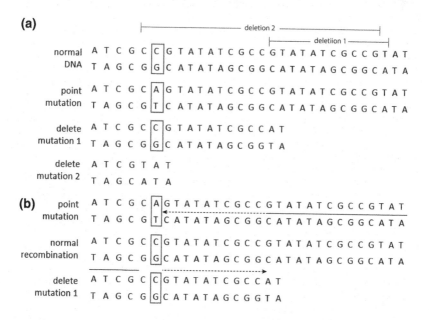

Fig. 5.8 Print-based sequence data can be cumbersome to examine. Created based on Yanofsky (1967, p. 88) (color figure online)

development of more efficient, powerful, and versatile sequence visualization tools (see Rebeiz and Posakony 2004; Sallaberry et al. 2011; National Center for Biotechnology Information 2016c).

But these tools and associated efforts, as may be imagined, are not focused on the needs of the publics. True, many of the tools, certainly those on NCBI, which is a US government agency, are "publicly available." That, however, is not to be taken to mean "accessible to the publics." While anyone can access the NCBI website and use its online databases and visualization tools without even a sign-in requirement, the 50 or so available choices with unfamiliar names easily overwhelm public users. Even with a chosen tool, its purpose, function, query method, and interface are not exactly what one would find instantly enlightening. Figure 5.9, for example, shows the NCBI Align Sequence Nucleotide Basic Local Alignment Search Tool (BLAST), which was

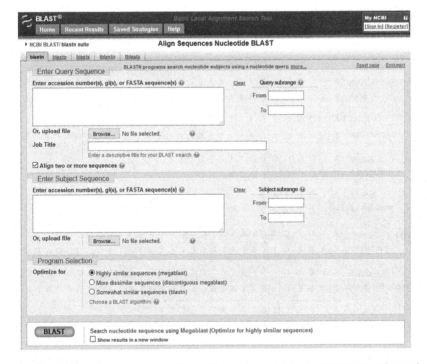

Fig. 5.9 NCBI Align Sequence Nucleotide Basic Local Alignment Search Tool (BLAST) (National Center for Biotechnology Information 2016a)

used to derive Fig. 5.7 shown above. Without knowing such jargon as "accession number," "gi," and "FASTA sequence" or how to obtain such information, one cannot get past this entry page to start a query,[1] let alone explore any visualization.

Just as the Human Genome Project, decades of development in sequencing technologies and digital tools failed to lead to complex code images for the publics. Indeed, as research and technology advanced and as scientists know more about the genetic code than they ever did, meaningful images about the code actually receded from the public view, replaced by exciting and ambiguous celebrations of it.

Third-Generation Sequencing: Public-Facing Visualization

If sequence data are too enormous to be presented in print with clarity and digital databases and tools are too complex for "lay" understanding, have we reached the end of the decades-long effort to illustrate the genetic code for public readers? Given the mysterious and fragmented code representations of late, such seems to have been a conclusion reached by scientists and science communicators. This conclusion, to some extent, reflects the necessary reality with scientific visualization: as subject matters and related visualization tools grow in complexity and embed more disciplinary information and technical know-hows, presenting the big picture of that information to an external audience becomes increasingly difficult. On the other hand, such a conclusion and reality are only necessary when researchers, practitioners, and decision makers in sciences, science education, and science visualization dedicate their efforts and budgets to assisting trained professionals while ignoring the need of the publics (more about this see Chap. 7). Without public-facing tools to unpack the genetic code, there is no tradition or evidence that publics *will* appreciate complex code images, nor is there research or examples on how to attempt such images. Representations that reduce the code to a handful of letters or elusive abstraction thus become convenient choices.

But if we believe that popular communication should do more than just inspire awe and appreciation, then the current reality is not acceptable. This is especially the case given the emerging third-generation sequencing. In 2011, the National Institutes of Health funded 14 million dollars to develop "third-generation," "everyday-use" DNA sequence technologies; the grant aims to further and significantly reduce current sequencing cost to less than $1000 per human genome so DNA sequencing can become a routine and rapid screening procedure performed by health care providers and researchers. When this becomes a reality, when obtaining our DNA sequence is as easy as obtaining a blood test report,[2] it seems relevant, essential even, that publics be able to examine, compare, and contemplate that information beyond its metaphorical façade.[3]

Digital visualization tools are certainly a promising means to reach this goal. Not only can they, as mentioned above, offer flexible and lucid presentation, they provide opportunities for user customization and interaction, which helps to promote deeper understanding and engagement (Tversky et al. 2002; Evans and Gibbons 2007). By this, I am not suggesting that publics be educated in using NCBI tools and the like. Designed to facilitate specialized research, these tools cannot meet the needs of novice users. What I suggest, then, are compact applications that do not require extensive biochemical training, are easily accessible via the Internet, and are involving without being overwhelming.

Very little effort currently exists to develop such public-facing tools. The Personal Genome Project—a global network that invites publics to share their genomic data for open access and analysis—offers a reporting tool. But the tool amounts to little more than large tables of gene variants and clinical details, which, as Shaer et al. (2015) found, are overwhelming and cumbersome to public audiences. That said, some tools that were originally developed for researchers or students may point to design possibilities. For example, widgets like SnipViz (Fig. 5.10), which offers a lightweight design and accessible functions such as base selection, comparison, and variation highlights (Jaschob et al. 2014), afford publics the ability to explore sequence data without the hindrance of overwhelming choices and dense terminologies. As another example, G-nome Surfer Pro, which was designed to help

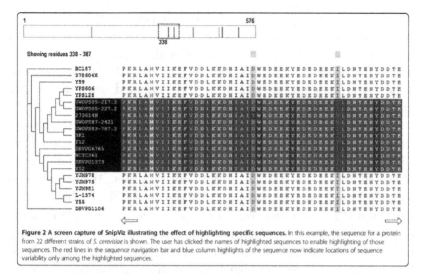

Figure 2 A screen capture of SnipViz illustrating the effect of highlighting specific sequences. In this example, the sequence for a protein from 22 different strains of *S. cerevisiae* is shown. The user has clicked the names of highlighted sequences to enable highlighting of those sequences. The red lines in the sequence navigation bar and blue column highlights of the sequence now indicate locations of sequence variability only among the highlighted sequences.

Fig. 5.10 Screen capture of SnipViz, a web widget for displaying gene and protein sequences (Jaschob et al. 2014, p. 4) (color figure online)

students explore prokaryotic genomes, is promising with its conscious effort to lower the threshold of bioinformatics tools and to provide an integrated user experience (Shaer et al. 2013). With the Surfer, users can explore sequence data in multiple, interrelated views: a whole-genome wheel, individual sequence slices, and relevant publications, all via touch-based interactions (Shaer et al. 2013).

Certainly, to ensure that any such tools are actually usable and useful to public readers, a range of questions needs to be explored. For example, what kinds of sequence data would public users be interested in exploring? Would they want to compare sequence data of different human populations or sequence data of humans and other organisms? How about normal versus disease-predisposing sequences? What kinds of communication contexts would they like to situate the data? What user interfaces would they prefer? And what user interactions would they desire? We currently have very few answers to such questions. Shaer et al. (2015) is the only publication found in this study that offers concrete advice. Based on their study of personal genomic data reporting, Shaer et al. (2015) recommended displays with color-coded visual summaries, glossaries (but reduced terminologies), integrated resources, and sharable information, among other considerations. More research from science communication scholars, genetics researchers and educators, and computer specialists is needed to develop digital presentations of the genetic code that are genuinely public-oriented, not just ones that exist in the public domain.

As we pursue such tools, print-based (or otherwise static) images will continue to play important roles in popular communication in the foreseeable future. It is thus useful to consider how we can build upon those "older" efforts to illustrate the genetic code on paper. As discussed earlier, such images face the challenge of visual clarity. While this is inevitable given the static print media and the complexity of sequence data, it *can* be negotiated by applying information design theories. Regarding the visually confusing Fig. 5.8, for example, the Gestalt principle of proximity may be used to signal related visual elements by putting them close together and to signal unrelated elements by positioning them apart, all to facilitate visual comparison. Figure 5.11 shows how this may be applied to part (a) of the original image. In Fig. 5.11, the normal DNA sequence is listed three times (as opposed to only once in the original) so it can be placed close to each mutation and facilitate comparison. Moreover, the distance that used to exist between the two

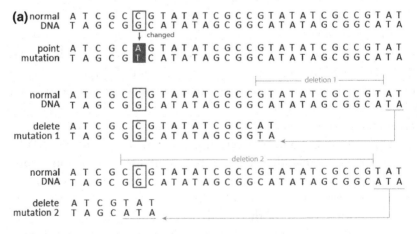

Fig. 5.11 Print-based sequence visuals can be re-approached to enhance clarity (color figure online)

strands of one same sequence is eliminated so the strands do not look like unrelated visual elements to be compared. In addition to proximity, Fig. 5.11 uses notation, shading, underlining, and arrows to increase visual contrast and clarity.

When sequence data are integrated into procedural and schematic illustrations, information design theories can similarly apply: for example, using strong color contrast and luminance contrast to emphasize base letters; minimizing non-essential visual elements such as swirling DNA strands; or, in general, reducing illustrations' iconic complexity (more about this see Chap. 4).

Last and probably most excitingly, a platform borrowed from the entertainment industry, namely gaming, has been used to introduce publics to the genetic code and may—with reconceptualization—prove to be a promising approach. These games are more precisely called games with a purpose (GWAP). Proposed by von Ahn in the mid-2000s, GWAP are designed to harness the conceptual and perceptual powers of individual human players and connect them via the Internet to act as a superior processor and perform tasks today's computers cannot or cannot perform well (von Ahn 2006). The more people who play a game, the more collective power can be harnessed and the more problems can be solved. To date, GWAP have been used to involve public players in a variety of disciplines and tasks: from describing online images for visually impaired users to correcting digital recognitions of historic texts.

In the field of genetics and with regard to the genetic code, one well-known GWAP is Phylo, which was developed in 2010 by the McGill Centre for Bioinformatics. Phylo is dubbed by popular media "a game that helps map genetic codes" (Hickey 2010) and likewise claimed by its developers as a game that helps "decipher our genetic code" (Phylo's Blog 2016). The task the game actually tackles is multiple sequence alignment: to align analogous sequences from different species so researchers can examine how sequences preserve or evolve, infer the functions of sequence regions, and examine their roles in genetic disorders (Kawrykow et al. 2012). Conventionally, this task is performed by complex computer algorithms, which are, however, prone to imperfections; Phylo developers, then, hope to harness humans' superior visual recognition ability to improve those alignments (Kawrykow et al. 2012).

To do so, the game displays DNA sequences in four colorful tiles that represent the four DNA bases of ATCG; players "slide" the tiles in multiple sequences for the goal of matching as many same-colored tiles as they can with minimum gaps (see Fig. 5.12). Performance is scored in real time based on optimal matching. Once players match or beat the existing score reached by the computer algorithm, they are deemed to have solved a puzzle and can advance to the next level. The solved sequences are automatically sent back to a server, evaluated, and applied (Kawrykow et al. 2012). The game proved to be an instant hit: within the first day of launching, it had attracted 1500 people who solved 7000 puzzles (Chung 2010). Within 7 months, 350,000 solutions were submitted by more than 12,000 registered users, and these solutions were found to help improve up to 70% of the alignment sequences played (Kawrykow et al. 2012).

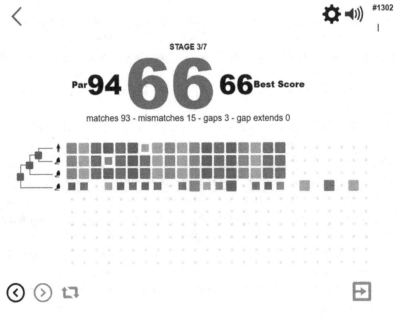

Fig. 5.12 Screen capture of a Phylo game (Phylo 2016) (color figure online)

A more recent, though lesser-known, GWAP that pertains to the genetic code is Fraxinus.[4] Launched in 2013, Fraxinus operates on roughly the same idea as Phylo: using human players to match sequence alignments. In this case, the sequences come from ash trees and a fungus (*Chalara fraxinea*) that infects the tree. With better alignments, researchers hope to identify sequences that make the fungus effective at infecting ash trees as well as sequences that make ash trees susceptible or resistant (Williams 2014). Fraxinus' interface and game setup are also similar to those of Phylo, though instead of squares, it uses leaf-shaped tiles to symbolize its subject matter. Similarly attractive, within 10 weeks of launching, Fraxinus had had more than 10,000 of its sequences examined (Williams 2014).

Certainly, the fact that these games were successful at attracting public players and generating valuable data is not necessarily evidence that they provided meaningful avenues for publics to explore the genetic code. In fact, both games "hide" the science behind them by removing any and all DNA base letters and providing no contextual information/explanation for the sequences in question. Phylo only chose to indicate which disease category a sequence is associated with. This information is supposed to make the players feel good about contributing to biomedical research (Kawrykow et al. 2012). Because of these design choices, the games look much like a Bejeweled game, a Tetris game, "an abstract puzzle game, with colorful shapes and jazzy music" (Grossman 2010). The purpose of this design, according to Phylo developers, is so that "people won't think about the biological problem, but just have fun and be entertained" (Grossman 2010) or, ultimately, so that the game can attract maximum players, not just science enthusiasts, to help solve its problems (Kawrykow et al. 2012). With such a design, one can be a master player of the games while gaining very little, if any, insight into the genetic code he or she is supposed to be helping decipher.

One may argue that as GWAP, these games are *not* supposed to inform publics or engage them beyond the fact of employing their work. What players gain for their contribution is entertainment, not knowledge of the problem they are solving (von Ahn and Dabbish 2008). In fact, von Ahn and Dabbish (2008) proposed that the players would have no interest in those problems anyway. This pragmatic stance, while it may be necessary to enlist citizens' help, may not be conscientious in light of current development in citizen science projects.

Gaining popularity in the last 5–10 years (though going back much further in history), citizen science projects are projects that employ everyday citizens who are not trained scientists to collect, categorize, transcribe, or analyze often large-scale scientific data (Bonney et al. 2014). One of the most well-established citizen science project formats, for example, is to employ bird-watchers worldwide to help gather data about birds' density, distribution, and migration, data that are tremendously difficult for individual researchers to gather on their own. Science-related GWAP, by enlisting the help of individuals in a virtual world, fall under the citizen science umbrella. Although citizen science projects as they were conceived easily emphasize the contribution of citizens to science rather than the other way around, scholars increasingly argue that these projects should not merely "use" publics as a research tool but should be attentive to public interest, allow participants to gain relevant insights, and engage publics in addressing questions of personal, local, and communal concerns (Bonney et al. 2014; Kaartinen et al. 2013).

This is not to suggest that games like Phylo and Fraxinus should contain a certain amount or kind of didactic lecture to address players' knowledge deficit about the genetic code. Rather, I propose that they be both emotionally and cognitively interesting; that they supply more than the kind of "fun" a Tetris game does; and that they afford, in Lewenstein's (2004) words, more realistic (even if messy) contexts, more meaningful (even if non-instructive) activities, and more social connections about and around science. To realize this goal again requires serious consideration and experiments by game designers, in conjunction with scientists and science communication scholars.

In this case, a way to broach the conversation may be via the field of education games. As scholars in this field have convincingly argued, education games, including science education games, have positive effects on players' motivation, attitude, collaborative participation, identity development, contextualized problem solving, as well as content knowledge (Squire et al. 2003; Freitas 2006; Chung and Wu 2011; Li and Tsai 2013; Gaydos 2015). Various design frameworks, approaches, principles, and pitfalls are currently being explored to best support these outcomes (Squire et al. 2003; Freitas 2006; Amory 2007). Although the conventional focus of education games is that players (i.e., learners and students) retain knowledge/concepts, scholars recognize that this focus is limited and recommend features such as authentic tasks, in-game discussions, and multi-player role-playing to help players understand the social–cultural situatedness of science and to develop communities of practice (Li and Tsai 2013). Such approaches may prove especially useful for reconceptualizing and redesigning public-facing GWAP.

Conclusion

Coined by a physicist based on speculation and echoed by two molecular biologists based on tantalizing evidence, the code metaphor had and continues to have a tremendous influence on the conception and communication of genetics. Even after decades of immense research, decoding life remains an elusive task—all the more so because what we know today allows us to realize how much we still don't know. The mysterious code metaphor not only sustained but deepened. In the realm of public communication, it has conjured up a range of images over time. In the mid-to-late twentieth century, on the heels of resolving the DNA structure and deciphering the DNA-to-protein mechanism, we saw rich images that attempted to visualize the code's technical detail and significance. But when it became clear that protein-coding DNA comprises only a small portion of life's secret and especially when sequencing technologies brought us more data than we know what to make of them, the genetic code assumed broader and renewed mystery as the entirety of biochemical information that gives rise to diverse life forms and functions. Following these developments, sophisticated digital databases and visualization tools appeared to help specialists decipher the broadened code. In popular communication, however, visual representations of the metaphor became less, not more, complex, though certainly far more exciting.

With its sheer size and complexity, the genetic code is difficult to visualize—for specialist audiences and public readers alike. Yet we clearly made and continue to make strides on one of these fronts, just not the other. Given the anticipated arrival of personalized genome sequencing, to make progress on the other front seems not only worthy but imperative. Considerable research and efforts are needed to understand public readers' information needs and preferences in this topic, to develop interactive digital tools that allow customized code visualization, to construct print images that offer convenient and accessible evidence, and even to design games that are both emotionally and cognitively engaging. This is not to say that we shall not be inspired by mannequins that sparkle with DNA bases, but that we need other visual representations to help us access, relate to, and contemplate those bases.

Notes

1. The "help" bubble that is present throughout the interface provides little help. For example, the bubble after the prompt "Enter accession number(s), gi(s), or FASTA sequence(s)," when clicked, reveals the following: "Enter query sequence(s) in the text area. It automatically determines the format of the input. To allow this feature, certain conventions are required with regard to the input of identifiers."
2. Limited numbers of personal genome sequencing are already available under $1000 (see Chap. 8).
3. While an expert advisor may be available to communicate testing results to patients, the sheer enormity of the data precludes the usual verbal communication; in addition, individuals will increasingly have independent access to genetic data (including clinical test results) and the ability to transport their data between services (Shaer et al. 2015).
4. *Fraxinus* is the Latin genus name for Ash.

References

Adams, J. (2008). DNA sequencing technologies. *Nature Education, 1*(1), 193.
Ades, J. (2005). ATCG's with silhouettes of people of varying heights. National Human Genome Research Institute. Retrieved July 19, 2016, from https://www.genome.gov/dmd/img.cfm?node=Photos/Graphics&id=85335.

Adleman, L. M. (1998, August). Computing with DNA. *Scientific American,* 54–61.

Amory, A. (2007). Game object model version II: A theoretical framework for educational game development. *Educational Technology Research and Development, 55*(1), 51–77. doi:10.1007/s11423-006-9001-x.

Bonney, R., Shirk, J. L., Phillips, T. B., Wiggins, A., Ballard, H. L., Miller-Rushing, A., & Parrish, J. (2014). Next steps for citizen science. *Science, 343*(6178), 1436–1437. doi:10.1126/science.1251554.

Cevallos, M. (2011, February 12). Alphabet of life: Searching for clues to the genetic code's origin. *Science News, 179*(4), 18–21.

Chung, E. (2010). Gamers tapped to crack DNA patterns. Retrieved July 19, 2016, from http://www.cbc.ca/news/technology/gamers-tapped-to-crack-dna-patterns-1.965281.

Chung, I., & Wu, Y. (2011). Digital educational games in science learning: A review of empirical research. In M. Chang, W. Y. Hwang, M. P. Chen, & W. Müller (Eds.), *Edutainment technologies: Educational games and virtual reality/augmented reality applications* (pp. 512–516). Berlin: Springer.

Cobb, M. (2015). *Life's greatest secret: The story of the race to crack the genetic code.* London: Profile Books Ltd.

Coleman, K. G., Poole, S. J., Weir, M. P., Soeller, W. C., & Kornberg, T. (1987). The invected gene of Drosophila: Sequence analysis and expression studies reveal a close kinship to the engrailed gene. *Genes & Development, 1*(1), 19–28.

Collins, F. (2011). *The language of life: DNA and the revolution in personalized medicine.* New York: Harper Perennial.

Crick, F. H. (1966). The genetic code: III. *Scientific American,* 55–62.

Dovichi, N., & Zhang, J. (2000). How capillary electrophoresis sequenced the human genome. *Angewandte Chemie International Edition, 39*(24), 4463–4468.

Drayna, D. (2005). Founder mutations. *Scientific American,* 78–85.

Dunham, I., Kundaje, A., Aldred, S., Collins, P., Davis, C., Doyle, F., … Birney, E. (2012). An integrated encyclopedia of DNA elements in the human genome. *Nature, 489*(7414), 57–74. doi:10.1038/nature11247.

Evans, C., & Gibbons, N. J. (2007). The interactivity effect in multimedia learning. *Computers & Education, 49*(4), 1147–1160. doi:10.1016/j.compedu.2006.01.008.

Fredholm, L. (2004). *Crack the code: How the code was cracked.* Retrieved April 14, 2016, from http://www.nobelprize.org/educational/medicine/gene-code/history.html.

Freeland, S. J., & Hurst, L. D. (2004, April). Evolution encoded. *Scientific American,* 84–91.

Freitas, S. (2006). *Learning in immersive worlds: A review of game-based learning.* London: Joint Information Systems Committee. Retrieved April 21, 2016, from http://www.jisc.ac.uk/media/documents/programmes/elearninginnovation/gamingreport_v3.pdf.

Gaydos, M. (2015). Seriously considering design in educational games. *Educational Researcher, 44*(9), 478–483. doi:10.3102/0013189X15621307.

Grada, A., & Weinbrecht, K. (2013). Next-generation sequencing: Methodology and application. *The Journal of Investigative Dermatology, 133*(8), 1–4.

Grossman, L. (2010, November 30). *Computer game makes you a genetic scientist.* Retrieved July 20, 2017, from http://www.wired.com/2010/11/phylo-game.

Harmon, A. (2008, March 4). Gene map becomes a luxury item. *The New York Times.* Retrieved April 18, 2016, from http://www.nytimes.com/2008/03/04/health/research/04geno.html.

Hickey, M. (2010, November 30). *Phylo, a game that helps map genetic code.* Retrieved May 3, 2016, from http://www.cnet.com/news/phylo-a-game-that-helps-map-genetic-code/.

Human Genome Mannequin. (2013). Retrieved August 10, 2016, from https://www.flickr.com/photos/greyloch/9121238998/.

Jaschob, D., Davis, T., & Riffle, M. (2014). SnipViz: A compact and lightweight web site widget for display and dissemination of multiple versions of gene and protein sequences. *BMC Research Notes, 7*(468), 1–6.

Jukes, T. H. (1963). The genetic code. *American Scientist, 51*(2), 227–245.

Kaartinen, R., Hardwick, B., & Roslin, T. (2013). Using citizen scientists to measure an ecosystem service nationwide. *Ecology, 94*(11), 2645–2652.

Kawrykow, A., Roumanis, G., Kam, A., Kwak, D., Leung, C., Wu, C., … Waldispühl, J. (2012). Phylo: A citizen science approach for improving multiple sequence alignment. *PLoS ONE, 7*(3), e31362. doi:10.1371/journal.pone.0031362.

Knudsen, S. (2005). Communicating novel and conventional scientific metaphors: A study of the development of the metaphor of genetic code. *Public Understanding of Science, 14*(4), 373–392.

Lewenstein, B. V. (2004, June 8). What does citizen science accomplish? Paper read at CNRS colloquium. *Paris, France.* Retrieved October 19, 2016, from https://ecommons.cornell.edu/handle/1813/37362.

Li, M., & Tsai, C. (2013). Game-based learning in science education: A review of relevant research. *Journal of Science Education and Technology, 22*(6), 877–898. doi:10.1007/s10956-013-9436-x.

Mattick, J. S., & Makunin, I. V. (2006). Non-coding RNA. *Human Molecular Genetics, 15*(01), R17–R29. doi:10.1093/hmg/ddl046.

Maxam, A., & Gilbert, W. (1973). The nucleotide sequence of the lac operator. *Proceedings of the National Academy of Sciences, 70*(12), 3581–3584.

Maxam, A. M., & Gilbert, W. (1977). A new method for sequencing DNA. *Proceedings of the National Academy of Sciences, 74*(2), 560–564.

Metzker, M. L. (2005). Emerging technologies in DNA sequencing. *Genome Research, 15*(12), 1767–1776.

Mitchelson, K. R., Hawkes, D. B., Turakulov, R., & Men, A. E. (2007). Overview: Developments in DNA sequencing. In K. R. Mitchelson (Ed.), *New high throughput technologies for DNA sequencing and genomics* (pp. 3–44). Oxford: Elseier.

Nassir, R., Kosoy, R., Tian, C., White, P. A., Butler, L. M., Silva, G., ... Seldin, M. F. (2009). An ancestry informative marker set for determining continental origin: Validation and extension using human genome diversity panels. *BMC Genetics, 10*(39), 1–13.

National Center for Biotechnology Information. (2016a). Align sequences nucleotide BLAST. Retrieved April 20, 2016, from http://goo.gl/YYfp5o.

National Center for Biotechnology Information. (2016b). Bagarius suchus cytochrome oxidase subunit I (COI) gene, partial cds; mitochondrial. Retrieved April 20, 2016, from https://goo.gl/fL9i9G.

National Center for Biotechnology Information. (2016c). Selected analysis tools. Retrieved April 20, 2016, from http://www.ncbi.nlm.nih.gov/home/analyze.shtml.

National Human Genome Research Institute. (2010). The Human Genome Project completion: Frequently asked questions. Retrieved April 28, 2016, from https://www.genome.gov/11006943.

National Human Genome Research Institute. (2011). *Comparative genomics*. Retrieved June 1, 2016, from https://www.genome.gov/11509542.

National Human Genome Research Institute. (2012). *An overview of the Human Genome Project*. Retrieved April 28, 2016, from http://www.genome.gov/12011239.

National Institutes of Health. (2011). NHGRI funds development of revolutionary DNA sequencing technologies. *NIH News*. Retrieved June 5, 2016, from http://www.genome.gov/27545118.

Nelkin, D., & Lindee, M. S. (2004). *The DNA mystique: The gene as a cultural icon*. Ann Arbor: University of Michigan Press.

Nerlich, B., Dingwall, R., & Clarke, D. D. (2002). The book of life: How the completion of the human genome project was revealed to the public. *Health: An Interdisciplinary Journal for the Social Study of Health, Illness and Medicine, 6*(4), 445–469. doi:10.1177/136345930200600403.

Petersen, A. (2001). Biofantasies: Genetics and medicine in the print news media. *Social Science & Medicine, 52*(8), 1255–1268. doi:10.1016/S0277-9536(00)00229-X.

Phylo. (2016). Retrieved December. 9, 2016, from http://phylo.cs.mcgill.ca/.

Phylo's Blog (2016). About. Retrieved April 22, 2016, from http://blogs.cs.mcgill.ca/phylo/sample-page/.

Pray, L. A. (2008). Discovery of DNA structure and function: Watson and Crick. *Nature Education, 1*(1), 100.

Public Broadcasting Service. (2001). The common genetic code. Retrieved March 23, 2016, from http://www.pbs.org/wgbh/evolution/library/04/4/l_044_02.html.

Rebeiz, M., & Posakony, J. W. (2004). GenePalette: A universal software tool for genome sequence visualization and analysis. *Developmental Biology, 271*(2), 431–438. doi:10.1016/j.ydbio.2004.04.011.

Rosen, M. (2015, December 12). What's in a face? *Science News, 188*(12), 24–26.

Sallaberry, A., Pecheur, N., Bringay, S., Roche, M., & Teisseire, M. (2011). Sequential patterns mining and gene sequence visualization to discover novelty from microarray data. *Journal of Biomedical Informatics, 44*(5), 760–774. doi:10.1016/j.jbi.2011.04.002.

Sanger, F., Nicklen, S., & Coulson, A. R. (1977). DNA sequencing with chain-terminating inhibitors. *Proceedings of the National Academy of Sciences of the United States of America, 74*(12), 5463–5467.

Satter, R. G. (2010, August 30). Scientists: We've cracked wheat's genetic code. *USA Today.* Retrieved March 20, 2016, from http://usatoday30.usatoday.com/tech/science/2010-08-27-wheat-genome_N.htm.

Schonrock, N., & Götz, J. (2012). Decoding the non-coding RNAs in Alzheimer's disease. *Cellular and Molecular Life Sciences, 69*(21), 3543–3559.

Schrödinger, E. (1944). *What is life? The physical aspect of the living cell.* Retrieved April 14, 2016, from http://whatislife.stanford.edu/LoCo_files/What-is-Life.pdf.

Shaer, O., Mazalek, A., Ullmer, B., & Konkel, M. (2013). From big data to insights: Opportunities and challenges for TEI in genomics. In *Proceedings of the 7th International Conference Tangible, Embedded and Embodied Interaction* (pp. 109–116). Barcelona, Spain. doi:10.1145/2460625.2460642.

Shaer, O., Nov, O., Okerlund, J., Balestra, M., Stowell, E., Ascher, L., ... Ball, M. (2015). Informing the design of direct-to-consumer interactive personal genomics reports. *Journal of Medical Internet Research, 17*(6), e146. doi:10.2196/jmir.4415.

Sinha, G. (2000, July). The power of knowing our genes. *Popular Science,* 55–57.

Skirble, R. (2013, June 27). Smithsonian explores genome revolution. *Voice of America.* Retrieved December 7, 2016, from http://www.voanews.com/a/smithsonian-genome-exhibit-unlocks-21-century-science-of-life/1690781.html.

Skloot, R. (2004, January). Putting the gene back in genealogy. *Popular Science,* 78–84, 100.

Smith, M. (1979). The first complete nucleotide sequencing of an organism's DNA. *American Scientist, 67*(1), 57–67.

Sosinsky, B. A. (2008). *Microsoft windows server 2008: Implementation and administration.* Indianapolis, IN: Wiley.

Spiegelman, S. (1964). Hybrid nucleic acids. *Scientific American,* 48–56.

Squire, K., Jenkins, H., Holland, W., Miller, H., O'Driscoll, A., Tan, K. P., & Todd, K. (2003). Design principles of next-generation digital gaming for education. *Educational Technology, 43*(5), 17–23.

Tejedor, J. R., & Valcarcel, J. (2010). Breaking the second genetic code. *Nature, 465*(7294), 45–46.

Tversky, B., Morrison, J. B., & Betrancourt, M. (2002). Animation: Can it facilitate? *International Journal of Human-Computer Studies, 57*(4), 247–262.

van Dijck, J. (1998). *Imagenation: Popular images of genetics.* New York: New York University Press.

von Ahn, L. (2006). Games with a purpose. *Computer, 39*(6), 92–94.

von Ahn, L., & Dabbish, L. (2008). Designing games with a purpose. *Communications of the ACM, 51*(8), 58–67.

Watson, J. D., & Crick, F. H. (1953). Genetical implications of the structure of deoxyribonucleic acid. *Nature, 171*(4361), 964–967.

Wetzel, R. (1980). Applications of recombinant DNA technology. *American Scientist,* 664–675.

Williams, S. (2014, Spring). Biology: A game for a crowd. *Biomedical Computation Review,* 4–6.

Yanofsky, C. (1967). Gene structure and protein structure. *Scientific American,* 80–94.

6

The Graph View: Navigating Big Data Science

Believed to be invented in the late eighteenth century by William Playfair, James Watt, and Johann Heinrich Lambert, Cartesian graphs became popular in the late nineteenth century and grew to dominate scientific publications in the twentieth century (Brasseur 2003; Gross and Harmon 2013). Although individual graphs, say a bar graph and a scatter plot, may look distinct in appearance, they rely on the same Cartesian coordinate system (i.e., the horizontal X-axis and the vertical Y-axis) to map quantitative data for trends and patterns—hence the common name Cartesian graphs. The rise of this visual genre in science, according to Gross and Harmon (2013), has to do with "a major shift in the emphasis of scientific practices from the gathering of observations to the generation and analysis of data, measured or calculated" (p. 5). This paradigm shift is apparent in the field of genetics, where data analysis had a humble beginning amidst classical observation studies but rose to center stage with the emergence of the DNA age. Fast forward another 50 years, and contemporary genetics has become a frequently invoked example of big data science with its exploration of DNA bases in the billions. Professional journal publications, accordingly, reply heavily on Cartesian graphs to provide comparisons, contrasts, and correlations.

© The Author(s) 2017
H. Yu, *Communicating Genetics*,
DOI 10.1057/978-1-137-58779-4_6

In popular communication, however, the story becomes considerably muddier. While the emergence of modern genetics saw a sharp increase in the use of Cartesian graphs, a growing awareness of public readers' information needs also coincided with the genre's decline—as well as its new development. Contributing to these shifts, as this chapter shows, are scientific as well as extra-scientific factors, each with layered complexities and defying one-sided conclusions. As a unique (some may say only) genre to transform quantitative data into visual patterns, Cartesian graphs are not only evidence of scientific inquiries, but the embodiment of modern science itself. Examining how they have been used in the popular communication of genetics, this chapter considers the role this visual genre plays (and may play) in the public uptake of genetics.

Counting and Measuring: Relatable Data from Classical Genetics

The quantitative foundation of modern genetics, once again, traces back to Gregor Mendel, whose well-known garden pea experiments established the Mendel's Law of Inheritance. What is less known—or at least less emphasized—is the quantitative nature of Mendel's Law and how Mendel arrived at his numerical conclusions. According to Mendel's Law, the dominant and recessive versions of a gene (the alleles) work in tandem to influence inheritance. When two alleles (say a dominant allele C for yellow-colored peas and a recessive allele c for green-colored peas) are passed down from parents to descendants, the alleles would randomly combine, resulting in a statistically equal distribution of 1/4 CC, 1/4 Cc, 1/4 cC, and 1/4 cc. Because the dominant allele C masks the recessive c, 3/4 of the descendants (the CC, Cc, and cC) would exhibit the dominant yellow color, and 1/4 (the cc) would exhibit the recessive green color, giving rise to a 3:1 ratio.

Mendel arrived at these conclusions through a modest method: by physically counting the thousands upon thousands of pea plants, pods, and peas in the various characteristics that he experimented with. In one experiment, Mendel cross-pollinated plants bearing round peas with those bearing wrinkled peas and obtained an entire first generation (F1) of round-pea plants, which suggests that roundness is a dominant trait (Griffith et al. 2000). Mendel then self-pollinated F1 plants to each other and grew a second generation (F2). In F2 plants, he counted 5474 plants with round peas and 1850 with wrinkled peas, a 3:1 ratio (Griffith et al. 2000). Pressing on, Mendel self-pollinated F2 plants and obtained the same ratio in their descendants, the F3 plants. It is through such repeated counting and accounting—using altogether 17,290 F3 plants in all experiments (Dyson 2015)—that Mendel was able to accurately establish his law of inheritance.

Other experimenters before Mendel had actually reached similar results but did not count and quantify (and thereby legitimate) their findings the way Mendel did (Griffith et al. 2000). Darwin came quite close: He conducted the same experiments using snapdragon plants and counted his results too, but he used a sample too small to yield statistically meaningful conclusions (Dyson 2015). It was Mendel's decision to quantify and to sample extensively that allowed him, in Griffith et al.'s (2000) words, to mark the birth of modern genetics. Mendel published snippets of his data (Fig. 6.1), though there is no evidence that he actually graphed any of them. Nonetheless, his practice set the stage for modern genetics' growing enthusiasm in quantitative analysis, which brought about Cartesian graphs in both professional and popular publications.

	1. Versuch.			2. Versuch.	
	Gestalt der Samen.			Färbung des Albumens.	
Pflanze	rund	kantig		gelb	grün
1	45	12		25	11
2	27	8		32	7
3	24	7		14	5
4	19	10		70	27
5	32	11		24	13
6	26	6		20	6
7	88	24		32	13
8	22	10		44	9
9	28	6		50	14
10	25	7		44	18

Fig. 6.1 Mendel's quantitative accounts of pea seeds' forms and colors (Mendel 1866, p. 13). The text in part 1 reads, from top to bottom and left to right, *Experiment 1, Form of Seed, Plants, Round,* and *Wrinkled.* The text in part 2 reads *Experiment 2, Color of Seeds, Yellow,* and *Green* (color figure online)

In popular communication, such graphs appeared in the early twentieth century. These early graphs show an obvious relationship to, if not direct influence by, Mendel's work. Figure 6.2, for example, graphs the findings of Dutch botanist Hugo de Vries, who examined 97 *Chrysanthemum segetum* (corn marigold) plants and counted the number of flowerets in each of their flower heads (Borel 1914). As the graph shows, a range of floweret numbers (everywhere from 12 to 22) was found and marked on the X-axis. Vertical lines above these numbers represent how many plants were found to bear those numbers of flowerets.[1] As it turned out, having 21 flowerets was the most typical.

Other early graphs, reflecting the focus of classical genetics, similarly plot directly observable, if not necessarily counted, data: from the number of different mice strains that survived diseases (Gowen 1935) to bone sizes and organ weights of normal versus dwarf chicken embryos (Landauer 1940). Unlike Fig. 6.2, many of these graphs make pointed arguments beyond simple data collection. Figure 6.3, for example, demonstrates Francis Galton's Law of Filial Regression, or the so-called tendency to mediocrity, which states that inheritance will tend to even out extreme variations toward the average. In this graph, a solid slope maps the heights of individual parents, each identified as a small circle; a dotted curve maps the heights of their respective children. As shown by the arrows, taller parents would beget shorter children, and shorter parents, taller children, with the results tending toward the mean height marked by a horizontal dotted line.

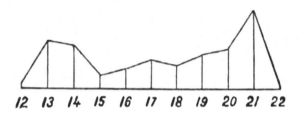

Fig. 10.—Curve representing the number of flower sets of chrysanthemum segetum.

Fig. 6.2 Number of flowerets among *Chrysanthemum segetum* plants (Borel 1914, p. 404)

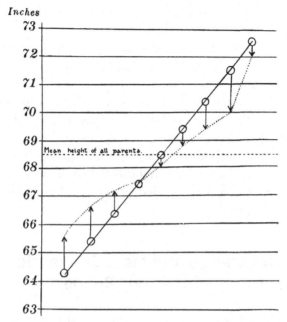

FIG. 49. DIAGRAM TO ILLUSTRATE THREE KINDS OF INHERITANCE DESCRIBED BY
GALTON. (After Walter.)

Fig. 6.3   Francis Galton's Law of Filial Regression, or the tendency to mediocrity
(Conklin 1914, p. 324)

These graphs from the classical genetic era are arguably easy for everyday readers to understand, at least based on the standard psychological
understanding of how people interpret graphs. According to this understanding, graph interpretation is a three-step process: First, readers perceive the graph at the syntactic level, e.g., locating the axes and slopes;
second, readers store, in short-term memory, the image they perceived
and its semantic meaning, e.g., recognizing that the vertical axis measures parents' and children's heights; last, readers make sense of this
stored image and information by invoking relevant, long-term memory
and knowledge, e.g., realizing that extreme heights in parents are evened
out in their descendants (Kosslyn 1989). Given this process, everything
being equal, graphs that contain less syntactic details (e.g., fewer lines
or data points) are easier to comprehend because they generate less

visual output. More importantly, graphs that display familiar and relatable data will be easier to comprehend because they can be more easily incorporated into readers' short-term and long-term memories.

Both of these conditions are generally met by popular graphs from the classical genetic era. Not only are these images relatively simple at the syntactic level with limited data sets, they display concrete, everyday objects that are relatable to public readers. Regardless of the amount of formal training, readers can envision measuring and comparing people's heights or the sizes of animal embryos. These graphs, then, join early photographs (Chap. 2) and illustrations (Chap. 4) to make for accessible visual evidence, a situation that, unfortunately, quickly changed in the following decades.

Formal Knowledge and Field Conventions: Specialist Graphs for Public Readers

As genetic research entered the DNA age in the 1950s, the number of graphs used in popular communication increased sharply, a trend that was most obvious in the 1960s and 1970s. For example, in the 1960s, over 30% of the genetics images used in *American Scientist* were Cartesian graphs, compared to 8.5% in the previous 2 decades. These graphs, most of them line graphs, were used to compare or correlate cellular and molecular experiment results: for example, organisms' cell sizes and amount of cellular DNA (Commoner 1964), ultraviolet absorption rates of DNA and RNA (Goodenough and Levine 1970), and enzyme activity levels of normal and mutant fruit flies (Morse 1984).

What is immediately noticeable about graphs from this era is that they are or resemble the archetypal modern scientific graph. They employ field-specific measurements and labels such as "cell volume— μ^3" without definition (see Commoner 1964); they emphasize data precision through the marking of individual data points; and they overlook syntactic attraction—notably, colors are used modestly and generally for data differentiation rather than visual appeal. As a whole, these graphs come across as serious, intelligent, and reliable conveyers of factual information that should summon reader attention and comprehension precisely because they are serious, intelligent, and reliable.

Figure 6.4, used in an *American Scientist* article (Ayala 1974), provides a glimpse of these effects. In this case, researchers introduced a gene variant (i.e., an allele) named 94 into four fruit fly groups: two groups of the *D. equinoxialis* species (E1 and E2) and two groups of the *D. tropicalis* species (T1 and T2). The frequencies of the allele were then tracked in these groups for 200 days or about 10 generations. As the flies propagated, the frequency of 94 increased in the E1 and E2 populations but decreased in the T1 and T2 populations, trending, in both cases, toward the natural frequency of the allele in the "wild" population. According to the author, this is evidence that the evolution of alleles is not random but selective.

As shown, Fig. 6.4 records precise data points rather than "smooth" lines, does not shy away from using the "broken axis" convention, and relies on one single color beyond black. The use of error bars is particularly noteworthy, as readers are expected to recognize what these bars are and to understand, as the original caption puts it, that they "encompass one standard deviation on each side of the observed frequencies" and

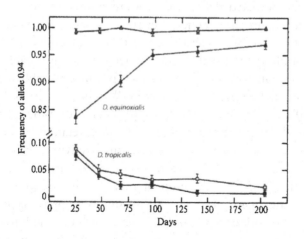

Fig. 6.4 Graph demonstrating the selective evolution of gene variants. Adapted by permission from Macmillan Publishers Ltd: Nature New Biology (Ayala and Anderson 1973, p. 700), copyright (1973) (color figure online)

that "the standard deviation was estimated from the binomial distribution" (Ayala 1974, p. 700). It is doubtful that these expectations could have been realistic, given non-specialists' general difficulty in interpreting standard deviation and its related statistical uncertainty (Delmas and Liu 2005; Gibson et al. 2013). According to the original caption, Fig. 6.4 is "adapted" from a graph the author had previously published in a research article in *Nature New Biology* (Ayala and Anderson 1973). Upon comparison, however, one realizes that the two graphs are identical except that the *American Scientist* version uses colored lines to replace the original solid and dashed black lines and that the names of the flies are written into the graph rather than in the caption. The fact that the author changed the graph, but only minimally, reflects a conscious (and unfounded) decision that public readers will (or should) grasp quantitative information from standard Cartesian graphs, just as specialists do.

It is important to note that in critiquing graphs like Fig. 6.4, I am not suggesting, as some scholars do, that public readers struggle with graphs because they supposedly lack high-level logical thinking structures or an ability to abstract from the visual concrete to the symbolic (Leinhardt et al. 1990; Berg and Phillips 1994). Such perspectives dovetail with the outdated deficit view toward public science communication and imply an elitist conclusion that quantitative analysis, and by extension the core of modern science, is beyond public grasp. To the contrary, studies show that public readers *are* quite capable of interpreting quantitative data if those data are put in a relatable context and familiar format (see Ancker et al. 2006; Waters et al. 2006; Smerecnik et al. 2010). A more meaningful way to broach scientific graphs, then, is to regard them as a semiotic object, the interpretation (as well as creation) of which is as much a social and cultural activity as it is a scientific and technical one. As Roth and McGinn (1997) argued, graphs "take their meaning from the situation of their use in communities where members share many of the same assumptions, preconceptions, and common sense notions" (p. 96). Or, in the words of "graph psychology," readers need to have relevant concepts and knowledge in their long-term memory before they can make sense of perceived visual patterns in a graph. The problem, of course, is that the mid-to-late twentieth-century genetic research had experienced an explosion in its field-specific knowledge and conventions, which are considerably formalized and removed from the earlier days' concrete reality and which mounted significant barriers to public access.

Figure 6.5 is a case in point, which was used to demonstrate that in *Chlamydomonas* (a type of algae), DNA exists not only in the usual cell nucleus, but also in the chloroplast (the cell unit that captures energy from sunlight through photosynthesis).

Upon first look, Fig. 6.5 seems simple enough with only two data lines and obvious data patterns, but closer examination shows that the opposite is true. To start, neither of the axis measurements ("fraction number" and "ultraviolet absorption") is a commonly recognizable

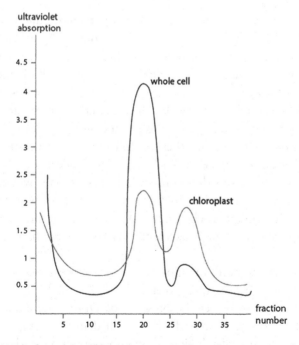

Fig. 6.5 DNA is found in *Chlamydomonas'* chloroplast. Recreated graph represents only general data trends. Created based on Sager (1964, p. 78) (color figure online)

concept. As such, how they correlate to each other to identify DNA remains moot. The original caption, quoted below, offers little help:

> CHLOROPLAST DNA, or deoxyribonucleic acid, is demonstrated by measuring the optical density in the ultraviolet of fractions of two DNA's: an extract from whole cells (black curve) and one from chloroplasts (color). A light fraction, about 5 percent of the whole-cell DNA, accounts for much of the chloroplast DNA: It must be localized in that organelle. (Sager 1964, p. 78)

The caption identifies what the two data lines are and offers a conclusion as to what the data demonstrate, but it provides no assistance for readers to interpret the graph toward that conclusion. In particular, the "5 percent" measurement mentioned in the caption seems apparently absent in the graph.

What, then, *does* Fig. 6.5 mean, and what knowledge and conventions does it assume of readers? As it turns out, there are at least four pieces of background information, including ones that are entirely absent from the depicted image. First, readers need a basic understanding of centrifugation. Put in simple terms, this is a process wherein a cell sample is spun at high speeds so cell materials can separate: heavier materials sediment toward the bottom, and lighter ones move toward the top. These materials can then be collected, one fraction at a time, from the bottom of the centrifugal tube. Each collection is known as a *fraction*. Earlier fractions thus contain heavier materials, while later ones contain lighter materials. It is by knowing this process, which is entirely absent from the graph itself, that readers can appreciate the "fraction number" measured by the X-axis and the reference to "light fraction" in the caption.

Second, readers need to know the basic process of how DNA is detected. DNA absorbs ultraviolet (UV) light, so, by measuring UV absorption, one can detect the presence of DNA. The amount of UV absorbed by DNA is expressed via "optical density," which is the ratio between the amount of UV light shone upon a sample and the amount of light that passes through the sample. A high ratio means significant UV light is absorbed by the sample and indicates the presence of DNA, and the higher the ratio, the more DNA is present. Only by knowing this process, which is again absent from the graph, can readers appreciate the "ultraviolet absorption" measured by the Y-axis.

With these understandings, a reader can finally start to process the context of the graph: a whole *Chlamydomonas* cell and its chloroplast are both spun at high speed; fractions are collected from each sample; these fractions then undergo UV tests to detect DNA. But even at this stage, how to process the data trends remains unclear. To do so, readers need to know the third convention: namely, the two data lines in Fig. 6.5 do not, as one might expect, indicate the presence of DNA. Instead, only the *peaks* of the lines do. This is because certain amounts of UV light will be absorbed by cell extracts regardless of DNA, so only when the absorption is significantly high can one confirm the presence of DNA. The whole-cell line has two obvious peaks, one around the heavier, 20th fraction and the other around the lighter, 27th fraction. The whole cell, therefore, has two types of DNA, one heavier and one lighter. The chloroplast line has two peaks at the same fractions, indicating that it, too, contains these two types of DNA.

Now comes the last hurdle: A reader is expected to be able to compare the amounts of different DNA in the two samples. Again, one does not, as one might expect, use the heights of the peaks for this comparison—because the Y-axis merely measures how much UV light is absorbed, not how much DNA is present. Specific formulas are needed to convert one measurement to the other. It is, however, possible to visually estimate the amounts of DNA by comparing the *sizes* of respective peaks. With the whole-cell line, the size of the 27th fraction peak is about 5% of the combined size of the two peaks (this is where the 5% in the caption came from); by contrast, with the chloroplast line, the size of the 27th peak is much larger. Because this lighter DNA accounts for a very small portion of the whole-cell DNA but a large portion of the chloroplast DNA, it is believed to be uniquely located in the chloroplast.

As this example shows, graphs with seemingly simple data patterns can embed a tremendous amount of background knowledge and various disciplinary conventions. These knowledge and conventions, bound up with the practice of formal scientific experiments, will be unfamiliar—and likely uninteresting—to public readers who do not have specialized training in molecular biology. As a result, there is very little chance that readers will find the graphs comprehensible, even if they are otherwise familiar with the graph genre. But could we, some might ask, see graphs like Fig. 6.5 as commendable efforts to present original experiment data for public access? To me, it is doubtful that there was any such conscious decision making: graphs from the mid-to-late twentieth century appear to simply be modeled after those prepared for professional journal publication; had their creators given some—indeed, any—consideration of the public audience, it should have been apparent, as the above extensive "walk-through" shows, that any such "public access" is unlikely to happen.

Aside from a complete lack of audience consideration, it is tempting to attribute, at least partly, the frequent use of graphs like Fig. 6.5 in the popular communication of this era to extra-scientific factors and social, cultural contexts. Grounded as they are in quantitative rationality and parsimonious analysis, Cartesian graphs are not only displays of specific scientific data but also the embodiment of modern science itself. They are credited the "first visual form with genuine heuristic potential," capable of detecting "changes in data that would not otherwise be apparent," "representing law-like relationships as correlations," and "allowing easy comparisons between theory and experimental data" (Gross and Harmon 2013, p. 5). The mere presence of graphs thus bespeaks validity and sophistication, and the then-emerging discipline of genetics would do well to resort to this genre for disciplinary status as well as, in this case, public acceptance. As Tal and Wansink (2014) showed, when scientific/medical claims are presented along with graphs, consumers perceive such claims to be more convincing. Moreover, whether one can interpret the graphs is beside the point. In fact, seemingly plausible (which is a given with Cartesian graphs' embodiment of science) but inaccessible information prevents readers from dismissing it and encourages them to accept its inferred informativeness (Haard et al. 2004; Weisberg et al. 2008; Tal and Wansink 2014). If so, graphs in these earlier days of popular genetics communication helped, intentionally or otherwise, cultivate a sense of uninformed public appreciation and trust.

Attraction and Adaptation: Graphing on Publics' "Own Terms"(?)

Compared with the mid-to-late twentieth-century popular genetics graphs, those published since the turn of the twenty-first century are becoming more relatable and accessible to public readers. On the one hand, this change has to do with the disciplinary shift in genetic research: from "bench" experiments to what Condit (1999) called the age of genetic medicine. With this shift, graphs can zero in on the single most engaging topic for the publics: personal heath—and the promised genetic solution. Often in the form of a bar or simple line graph, these images compare the genetic profiles of organisms and individuals, reveal the correlations between genetic profiles and disease conditions, and examine the prospects of high-risk populations. In O'Brien and Dean (1997), for example, one witnesses in a line graph the significant difference between HIV-infected and non-infected individuals' CCR5 gene profile[2]; following that, another graph shows how CCR5 gene compositions correlate with the length of time it takes for HIV infection to progress to AIDS.

In addition to this disciplinary shift, there are apparent efforts from graph creators to inform, engage, and even attract public readers—efforts that correlate with the contemporary focus on public participation and engagement in science. These efforts take a number of approaches, which may well be combined in practice but are separately discussed below to clarify their respective complications and implications.

Syntactic Attraction

Compared with other visual genres discussed in this book, Cartesian graphs are not conducive to creating visual and affective appeals. This is because, as Brasseur (2003) put it, they are a stable visual genre with standard conventions, namely, the Cartesian coordinating and mapping system. The backbone of this system, the axes and grid, is perceptually austere; as for a graph's visual pattern, it is subject to the numerical coordinates of the data, not design or aesthetic considerations. What is more, graphs in their common forms such as lines, bars, and pies are familiar to public readers. Used in school, at work, and in magazines, newspapers, and advertisements, they are "a pervasive part of our environment" (Kosslyn 2006, p. 3). As such, they offer little visual novelty or excitement.

That said, local attempts at visual attraction are not impossible. One easiest way to do so, as may be imagined, is via color. Although graphs from earlier decades already used color, the usage was moderate: The number of colors included in a graph was limited, and the color was used primarily for data coding. In recent graphs, however, color is used far more liberally and frequently for what Pozzer and Roth (2003) would call decorative functions. Background color, for example, is applied to cover the coordinate system (i.e., the area enclosed within the X and Y axes), which produces a more eye-catching display but, as Kosslyn (1989) pointed out, serves no role in communicating quantitative information.[3] In addition, when multiple data sets are graphed, rather than using black/gray for some sets to reduce the need for true colors (as earlier graphs do), creators favor color coding all data sets.

In addition to color, the dimension is another "easy" syntactic element to add to the visual attraction of a graph. Figure 6.6 is one example,

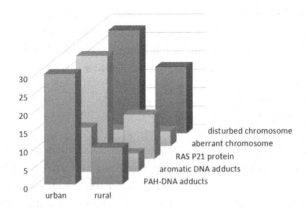

Fig. 6.6 Three-dimensional bar graph compares urban and rural residents' DNA. Recreated graph represents only general data trends. Created based on Perera (1996, p. 60) (color figure online)

which compares DNA profiles of residents from polluted urban areas and those from rural areas. As it turned out, urban residents' DNA contains higher levels of adducts (complexes formed when carcinogens combine with DNA or proteins), abnormal chromosomes, and RAS P21 protein (which is implicated in several cancers).

If increased color use does not necessarily interfere with graphs' ability to illustrate or explain data,[4] the same may not be said about three-dimensional graphs like Fig. 6.6. Although different findings do exist, it is generally accepted that three-dimensional graphs are prone to observational issues (Kosslyn 2006). In the case of Fig. 6.6, it is ambiguous which line in the top plane of a bar should be used to measure the bar's height. In addition, dimensional columns are obscuring data sets or making them difficult to measure. More generally, because graphed data are often not of a three-dimensional nature and because our visual system does not estimate volume very gracefully, graphs like Fig. 6.6 do not facilitate quantitative information processing (Kosslyn 2006; Amare and Manning 2013). What they *are* able to achieve, though, is create a sense of professional ethos—thanks to the graphs' apparent visual sophistication, an implied effort on the part of their creators for presentation effect, and the creators' demonstrated visualization skill (Kosslyn 2006). In other words, the added dimension serves to evoke the same inferred informativeness and implicit public appreciation as inaccessible Cartesian graphs do.

Integration of Pictorial Elements

Since Cartesian graphs themselves do not accommodate much syntactic embellishment, creators are seeking ways to combine graphs with relevant and syntactically rich images. A simple way to do so is to physically juxtapose graphs with those images. For example, a graph that compares a drug and placebo's effect on preventing breast cancer may be presented side by side with a photograph of women who participated in the study (Jordan 1998). Or, a series of graphs that correlates cocaine use with gene activities may be presented in conjunction with a striking illustration of how cocaine structurally impacts DNA (Nestler 2011).

A more ingenious approach, though, is to integrate pictorial elements into the graph proper. Figure 6.7 offers one example. It examines the genetic pedigree of the coelacanth, a rare fish believed by some scholars to be most closely related to the fishes that first crawled onto land and became four-limbed vertebrates. Figure 6.7 is among the evidence that supports this theory. It compares the hemoglobin molecules[5] of several ancient fishes with that of a vertebrate, frog (more precisely, baby frogs, the tadpole), demonstrating the extent of their match. In

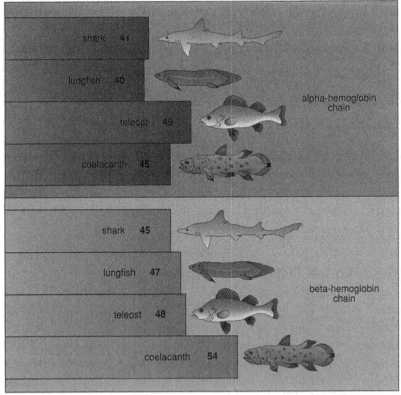

average amino-acid matches between fishes and tadpoles (percent)

Fig. 6.7 Integrating pictorial icons into a graph (Gorr and Kleinschmidt 1993, p. 80). Reprinted by permission of *American Scientist*, magazine of Sigma Xi, The Scientific Research Society (color figure online)

addition to the familiar bars with textual labels and quantitative data, the graph uses pictorial icons to illustrate the fishes. As the graph (lower half) shows, coelacanths and tadpoles have the highest amino acid match in the beta chain of the hemoglobin molecule, demonstrating their evolutionary relatedness. Although teleosts and tadpoles have the highest match in the alpha chain (top half of the graph), as the authors explained, the alpha chain experienced high and variable mutation and is not reliable evolutionary evidence. This pictorial icon visual strategy is used in various other instances, graphing, for example, the amount of non-protein-coding DNA in different organisms (Mattick 2004) or the height-to-weight ratios of different dog breeds due to genetic variations (Ostrander 2007).

Although pictorial elements such as those used in Fig. 6.7 do add surface-level interest and "liven up" otherwise standard graphs, they are not purely decorative. At the most basic level, they fulfill what Pozzer and Roth (2003) called an illustrative function and help depict potentially unfamiliar data (i.e., what is a lungfish?). Even when readers are familiar with the illustrated concepts, the images provide an immediately apparent visual context to situate the data; as studies show, doing so can help enhance data recall, retention, and interpretation (Bateman et al. 2010; Borgo et al. 2012). In addition, pictorial elements may well play a complementary function (Pozzer and Roth 2003) by providing information that is not present in the numbers. In the case of Fig. 6.7, readers are afforded the opportunity to compare the physical appearances of these animals and ponder their possible association with genetic profiles.

It should be noted that integrating pictorial elements into graphs is not necessarily a recent invention. Earlier graphs did the same when their data, occasionally, pertained to physical structure and appearance. For example, when comparing the shapes of DNA in different chemical environments, Bauer et al. (1980) drew those shapes directly into a graph. However, compared with these earlier examples, contemporary works use pictorial elements more deliberately even when the data in question do not have obvious visual dimensions. Doing so, then, represents an inventive choice to appeal to public readers, rather than an automatic method to illustrate data.

To some readers, such elective choices may sound suspicious, thanks in no small part to Edward Tufte's well-known work on chartjunk: graphs that are overflowing with non-data ink or redundant data ink (Tufte 2001, p. 107). It is true that pictorial elements, in so far as they represent the same textual labels given in a graph, are semantically redundant. But such elements are not cognitively or affectively redundant. As I argued above, they reinforce information via the visual communication channel, provide readers with an immediate data context, and add emotional interest to standard Cartesian graphs. Used in moderation, they represent a promising, if localized, attempt to engage public readers.

Adaptation of Specialist Graphs

Because graphs used in popular science communication frequently draw upon those published in professional journals, a more holistic—and thus arguably more genuine—way to engage publics is wholesale adaptations of those graphs. These adaptations are noteworthy in their intention—and, as I argue, their ability—to make specialized, quantitative evidence accessible and meaningful to public audiences. Particular adaptation methods vary, but heuristically, they address the multiple steps in graph processing, including readers' syntactic identification, semantic understanding, and pragmatic interpretation of graph elements. Syntactically, adaptation considers whether data plotted in an original graph may be changed to create a clearer visual pattern. Semantically, it considers what data and graphing formats are likely to be familiar to public readers. Pragmatically, it emphasizes telling relatable stories about data rather than assuming that data are inherently valuable.

Figure 6.8 shows a before-and-after adaptation example. The graphs examine the genetic profiles of centenarians (red) and regular controls (blue), comparing their likelihood of carrying disease-associated gene variants. The original graph (Fig. 6.8 top) examines 15 gene variants, some of which are associated with singular diseases such as Parkinson's diseases (PD), and some are associated with a group of related diseases such as personality disorders. All 15 variants are listed along the Y-axis, each appearing twice to represent the centenarian and the control data. The

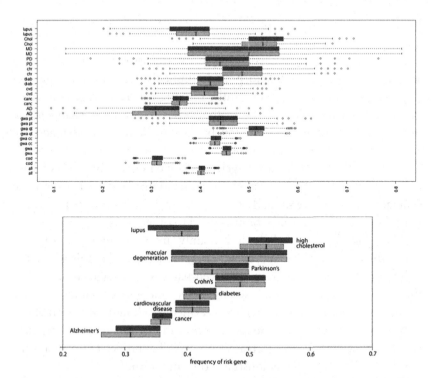

Fig. 6.8 *Top* Gene variants among centenarians and controls as shown in a box-plot (Sebastiani et al. 2012, p. 11). *Bottom* An adapted bar graph. Recreated graph represents only general data trends. Created based on Saey (2012, p. 20) (color figure online)

X-axis then measures the likelihood of corresponding gene variants in the two populations. Because the rates at which individuals carry a gene variant differ, the data for each population are a statistic range. The original graph, known as a boxplot, shows this range using set conventions: Horizontal boxes represent the middle 50% of the data (with a black line indicating the median), dashed lines to the two sides of the box show the 25% of the data above or below the boxed value, and circles to the sides of the lines stand for outliers that are out of the normal data range.

With all these details, the original graph has considerable syntactic complexity. More importantly, what these details signify is not self-evident to readers who are not already familiar with the boxplot format.

In adapting the graph for public readers in a Science News article (Saey 2012), the creator reduced the number of the gene variants to nine, thus removing some of the syntactic complexity; the associated disease names are also written out in full to provide immediate semantic recognition (Fig. 6.8 bottom). More importantly, the original data range is reduced to its essence—the middle 50%, or the "boxes"—effectively changing the unfamiliar boxplot into a common bar graph. With these changes, the adapted graph presents an obvious syntactic pattern: namely, the centenarians and the controls have largely overlapped data ranges. This recognition, in turn, affords a relatable interpretation: Centenarians and the general population are similarly likely to carry disease-associated genes, so their different life spans may be attributed to other factors.

With Fig. 6.8, the adaptation mainly involved reducing original data, which is a valuable approach, especially in cases like this where the reduction does not take away but accentuates the message embedded in the data. At the same time, if we agree that one goal of popular science communication is to make scientific evidence relevant and relatable to public readers, then graph adaptation should not be an automatic process of "showing less" but a creative process of genuine transformation.[6] Figure 6.9 more clearly demonstrates this intention.

The graphs in Fig. 6.9 compare how many copies of the protein-coding *KIAA1267* gene are found in Asian, European, and African

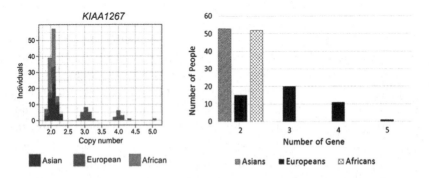

Fig. 6.9 *Left* Gene copy variations as shown in a stacked histogram. From Sudmant et al. (2010, p. 642). Reprinted with permission from AAAS. *Right* An adapted bar graph. Recreated graph represents only general data trends. Created based on Saey (2010, p. 20) (color figure online)

populations.[7] In both the original and adapted graphs, the X-axis lists possible copies of the *KIAA1267* gene, the Y-axis shows how many people carry a given number of the gene, and three colors/patterns respectively code the three populations. However, the original graph (Fig. 6.9 left), being a stacked histogram, is much more syntactically complex: the X-axis shows a scale with fractions; same-colored data for a single population appear in multiple units at each X-axis point; data "blocks" are irregular in shape and stacked, preventing an easy measurement.

These syntactic details reflect the data collection/analysis process used in the original study. Put somewhat simplistically, the data in question were obtained based on an analysis of 159 human genomes. But given the enormous DNA in the genome, the analysis was not a simple matter of "counting" the actual copies of the *KIAA1267* gene that exist. Rather, researchers randomly examined DNA fragments to obtain copy number data and then used statistical models to estimate the number of this gene in the three populations. Because of this, the original graph does not plot gene numbers in realistic terms (i.e., 1 copy, 2 copies, etc.) but in unrealistic, though statistically accurate, fractions.

These experiment and data analysis details were discarded when the graph was adapted in a Science News article (Saey 2010), where the data are rounded to idealized terms that proximate the reality (Fig. 6.9 right). For example, people who, statistically speaking, have 1.9 or 2 or 2.1 copies of the gene are all treated as having two copies. The graph thus becomes a familiar, grouped bar graph where the three populations are compared across four copy number possibilities. The syntactic patterns become obvious, as are their practical significance: namely, the three populations have varied genetic profiles; Asians and Africans generally have only two copies of the *KIAA1267* gene, whereas Europeans have anywhere between two and five copies.

It is true that with this change, finer details are still "lost," but that is not a result of the visual creator simply omitting part of the original data. It is, rather, a creative solution to solve the problem of the original graph demanding undue specialist knowledge from the publics. In this sense, data are not so much "lost" as they are re-situated within a different social context. This context has its own "common sense" rules (Bauer and Gaskell 2008) for "counting" gene copies, rules that go beyond the categorical demand for statistical accuracy and precision practiced in the formal scientific context. Such re-situation is not only inevitable but commendable when we balance the need to present meaningful data and the need to reach genuine public participation. It is as Kosslyn (1989) put it, in graphing, more information is not necessarily better; what matters is having the appropriate amount and type of information for an intended purpose. In popular communication, that purpose is *not* to afford verification, replication, or calculation of data but to communicate relatable trends and patterns (Brasseur 2003; Harmon and Gross 2013). Preoccupation with accuracy and precision may actually prevent readers from discerning those trends and patterns or making meaningful interpretations and decisions (Lloyd and Reyna 2001). Indeed, the previous decades' graphs have been quite accurate and precise, but their value, as seen before, is entirely lost beyond a select group of people.

Compared with the addition of surface-level attractions and the integration of pictorial elements, holistic adaptations are not concerned, or not just concerned, with a graph's local decorative, illustrative, or explanatory functions. They are focused instead on making a centrally important scientific visual genre accessible to publics. In that sense, they are more cognitively interesting, consciously engaging, and indeed empowering in the face of deficit assumptions that "lay" people lack "mental abilities" to process graphs.

Pictographs: Missed Opportunities

Whether by adding syntactic details, integrating pictorial elements, or creating a holistic adaptation, the graphing methods discussed so far assume the primacy of the Cartesian system and maintain its coordinate parameters. But this does not have to be the case once we start to entertain the possibility of pictographs.

Variously called icon arrays (Ancker et al. 2006; Galesic et al. 2009) or visual tables (Kosslyn 2006), pictographs employ icons such as human figures, everyday objects, and geometric shapes to express quantitative data. Generally speaking, in pictographs, a single icon is set to represent a numerical "base unit," and varying quantities are then expressed by displaying more or fewer icons. Unlike standard Cartesian graphs, pictographs have a much shorter history. Their origin is often attributed to the International System of Typographic Picture Education (Isotype), which was developed in the 1920s by Otto Neurath (1882–1945) and his colleagues at the Vienna Museum of Society and Economy.[8] At the time, Vienna was experiencing a period of municipal socialism (i.e., the "Red Vienna") and putting strong emphases on cultural and educational projects (Burke 2009). Born out of that context, Isotype was more than a graphing technique but also, or primarily, a cultural and political tool, a means to represent social facts (such as how municipal taxes were spent), to bring "dead statistics" to life for everyday citizens, and to explain how the citizens of Red Vienna "fitted into the world's complex of interconnections" (Burke 2009, pp. 211–212).

A classic Isotype example is shown in Fig. 6.10, which illustrates the number of people immigrated into and emigrated from various European and (North and South) American countries between 1920 and 1927. Country names are listed to the left of the graph, and populations are represented via luggage-carrying human figures (each figure represents 250,000 people, as noted in the footnote). The graph uses a vertical line as the population "break-even" point: Emigrants from a country are placed to the right of the line traveling away from the country; immigrants then pick up the "deficit" left by the emigrants and march left toward the country (immigrants are also drawn on a gray background). As a result, if there are more figures to the right of the "break-even" line, a country is losing population; otherwise, it is gaining population. And the number lost or gained can be reached by multiplying the human figures by 250,000.

Although the Isotype initiative resulted in a great many works from the 1920s to the 1950s and left a lasting social, cultural, and political influence across several continents (see Neurath and Ross 2009), in the field of data visualization, its impact has been modest, as contemporary research consistently favors standard Cartesian graphs and gives minimal coverage to pictographs (see, e.g., Jacques 1983; Tufte 2001; Kosslyn 2006). More damningly, when pictographs *are* mentioned, they are often explicitly or implicitly considered the "lesser" kind of graph.

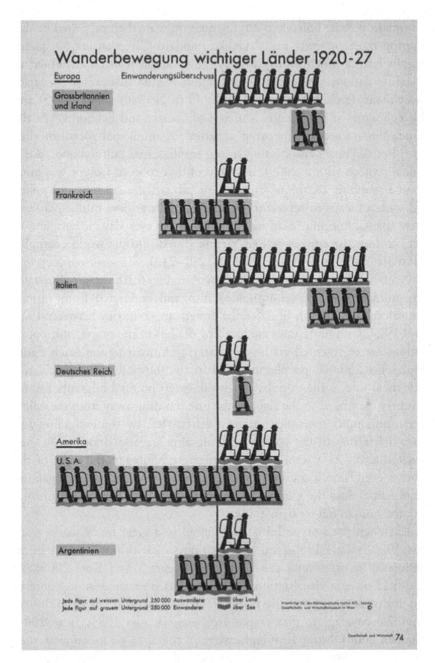

Fig. 6.10 Movement of people across countries, 1920–1927 (*Gesellschaft und Wirtschaft* 1930, p. 74) (color figure online)

As Kosslyn (2006) wrote, pictographs are "pared-down graphs" unable to present precise numbers (p. 45); worse, they easily result in "lying" graphs that are too busy with decorations to get the data right (Tufte 2001). This context helps to explain the lack of pictographs encountered in this study: Aiming to have their work and research be taken seriously, scientists and science communicators would not want to risk using an "inferior" graphing format.

But pictographs, I argue, are a missed opportunity. I say so not because, or not only because, the use of pictorial images can attract reader attention. While studies suggest this to be true (Bateman et al. 2010; Haroz et al. 2015), this attraction represents only an initial reader reaction and does not itself guarantee engaged reading. My argument, then, is based first and foremost on pictographs' ability to tell rich, intelligent, and memorable stories—thanks to their combined use of visual, numerical, and verbal cues. Figure 6.10, for example, invites data comparison at multiple levels. First, as already mentioned, it allows readers to tell, at a glance, whether a country is gaining or losing population and by how much. Second, it enables cross-country comparison: Without having to count the figures, readers can tell from the graph's overall pattern which countries gained more/most populations and which lost more/most. Third, the graph indicates the means by which people traveled: Those traveling by land have a brown, horizontal bar (symbolizing land) below their pictorial figures; correspondingly, those traveling by sea are signaled by a blue, wavy line. Trends in transportation methods within and between countries are thus visible. Moreover, Fig. 6.10 tells these stories in an engaging manner. Unlike standard Cartesian graphs, pictographs do not follow set coordinate conventions or output perceptually similar (even though statistically different) lines and bars. Rather, readers have to actively engage with each pictograph to figure out what icons or shapes are used, what they stand for, and how this shifting visual landscape translates to quantitative evidence.

Pictographs' these benefits have recently attracted the attention of scholars and practitioners in medical and risk communication. In these communication contexts, numerical data are frequently used to convey risk and efficacy, but these data are difficult for patients and medical staff alike to interpret and discuss (Elmore and Gigerenzer 2005;

Ancker et al. 2006; Garcia-Retamero et al. 2010). Pictographs, as studies and clinical practices suggest, present a promising solution (Edwards et al. 2002; Paling 2003; Galesic et al. 2009). Fagerlin et al. (2005), for example, found that pictographs allowed participants to actively use statistical information and make evidence-based medical treatment choices. Garcia-Retamero et al. (2010) suggested that pictographs help clarify data by emphasizing both the foreground, numerator information (i.e., how many people were cured by a treatment) and the background, denominator information (i.e., how many people received the treatment). Some studies also suggested that the pictorial icons used in pictographs make the displays "more meaningful, easier to understand, and easier to identify with" than standard graphs (Ancker et al. 2006, p. 612).

These benefits can be seen in Fig. 6.11, which shows the result of a hypothetical drug trial in which patients were put into a control group

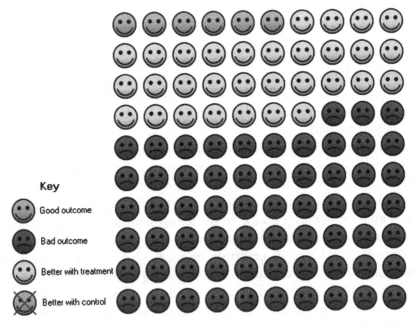

Fig. 6.11 Cates Plot of a hypothetical drug trial. Created using Visual Rx, a graphical display software developed by Christopher Cates (Cates 2008) (color figure online)

and an experimental group. The control group received a placebo, and 6% showed improved symptoms; the experimental group received the drug, and 37.3% showed improvement. Statistically speaking, then, 6% of the target population would show improvement with or without the drug, while another 31.3% would respond to the drug. By using face icons for differently affected individuals, Fig. 6.11 provides an intuitive and comprehensive display of these data[9] for readers to assess the drug's overall efficacy: Namely, it has no effect on the majority of the population (the red, "bad outcome" faces); it helped a significant minority (the yellow, "better with treatment" faces); of that minority, though, a fraction (the green, "good outcome" faces) would have gotten better without the drug. It is true that, in rounding numbers to pictorial units, Fig. 6.11 is not precise, but, as mentioned earlier, statistical precision is not a priority in popular communication—in the words of the Isotype team, "To remember simplified pictures is better than to forget accurate figures" (Burke 2009, p. 215).

In the popular communication of genetics, pictographs can be similarly used to express quantitative data about genetics-based disease risks and the promise of genetic medicine. Indeed, the few exceptions encountered in this study are precisely such works where stick figures are used to represent human populations and colors are applied to denote, for example, healthy versus diseased individuals (see insets in Gorman and Maron 2014). Direct-to-consumer genetic testing companies such as 23andMe[10] and deCODEme[11] also used this format in their testing reports. If, say, someone's genetic profile is such that in a comparable population of 100, 35 with the profile will develop Type 2 Diabetes, the report draws 100 stick figures and then color codes 35 of them.

Beyond these obvious applications, various other possibilities exist. Consider Fig. 6.12, which reflects the graphs used in a *Scientific American* article (Brown 2001) on the worldwide statistics of genetically modified (GM) crops. Part (a) of the figure highlights the percentages of GM crops among the total soybean, corn, cotton, and canola planted, part (b) compares the ways these GM crops were modified (herbicide tolerant, insect resistant, or both), part (c) breaks down the GM crops by crop type, and part (d) breaks them down by modified traits.

In many ways, Fig. 6.12 typifies the contemporary graphs encountered in this study with their liberal use of color and 3D effect. While not inaccessible, they are not particularly revealing or inviting either. A pictograph approach, by contrast, may create more engaged reading and, as Fig. 6.13 shows, integrate all of the original data in a single display to provide richer data comparison. Figure 6.13 highlights denominator information (total hectares of crops and total genetic modification) as well as numerator information (hectares of GM crops and specific modifications), thus revealing the various part-to-whole relationships in the data. Readers can assess, at a glance, how prevalent

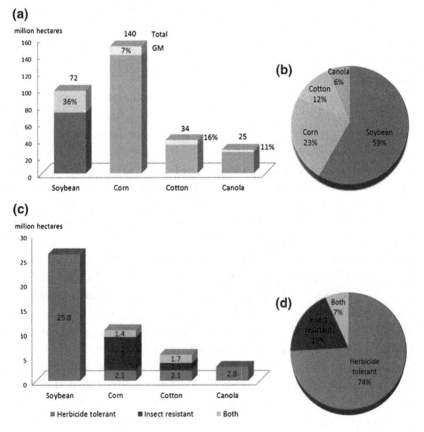

Fig. 6.12 Conventional graphs of worldwide GM statistics. *Source* Brown (2001) (color figure online)

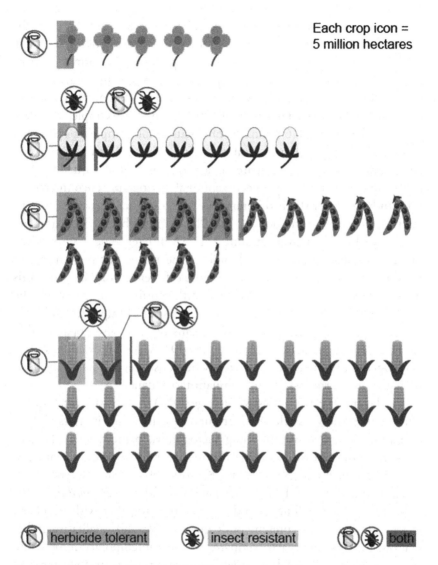

Each crop icon =
5 million hectares

Fig. 6.13 Pictograph of worldwide GM statistics (color figure online)

GM crops are both in general and within specific crops; how GM crops are modified across different crops; and which genetic modification is most/least common, again in general and for specific crops.

Certainly, many a research finding in genetics may not be conveniently represented by stick figures, smiley faces, or icons of everyday objects, but this should not preclude the pictographic format, as geometric shapes can be used in lieu of icons to represent data, a choice that yielded similarly positive outcomes in user experiments (Galesic et al. 2009; Garcia-Retamero et al. 2010). So, for example, the data shown earlier in Fig. 6.6 (amounts of carcinogenic complexes and DNA abnormalities in urban and rural residents) may be illustrated by using an arbitrary shape, say a square, as a single unit of abnormality. This is the approach taken by one exceptional pictograph encountered in this study (Fig. 6.14). In this rare case, the designer used dots to represent DNA bases (one dot for 500,000 pairs of bases). The dots are color-coded to denote the amount of DNA difference between modern humans, gorillas, chimpanzees, bonobos, and the extinct human species Denisovans: The darker the dot, the more difference in DNA. With this design, the graph shows, at a glance, that the gorilla is comparatively the most distinct from modern humans, whereas the chimpanzee and bonobo have similar DNA as modern humans do. It is also apparent that we and Denisovans share largely the same DNA. As this example shows, a pictographic approach can effectively demonstrate, in a relatively compact size, an enormous amount of quantitative data.

The layout Fig. 6.14 uses to arrange DNA dots is a fractal space-filling curve known as the Hilbert curve. In doing so, it starts to hint at a more inclusive way of using pictographs, namely, to go beyond rows upon rows of tiny icons and more generally engage the statistical through the pictorial (Neurath and Kinross 2009). Later Isotype works (such as those included in the *Compton's Pictured Encyclopedia*) exemplify this broader definition and use pictures, together with numbers, scales, and texts, to represent both quantitative and qualitative information. Such an approach compares to the contemporary infographic genre, which combines data visualization, illustration, and written text to tell meaningful stories (Randy 2013).

With this approach, we can more flexibly transform conventional Cartesian graphs to suit individual communication contexts. Consider Fig. 6.15, a graph from the 1960s that quantifies the structural changes during cell division. The X-axis documents the time (in minutes), and

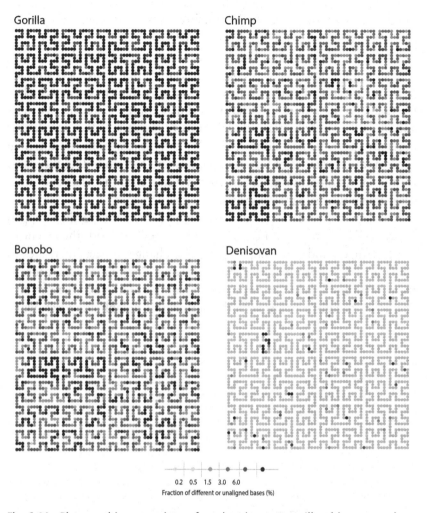

Fig. 6.14 Pictographic comparison of modern human, gorilla, chimpanzee, bonobo, and Denisovan genomes (Wong 2014, p. 100). ©Martin Krzywinski. Used with permission (color figure online)

the Y-axis measures the distances (in microns) between cellular components. Three distances are graphed: (1) the distance between a chromosome (or more precisely, its two chromatids) and a corresponding polar end, or pole, of the cell; (2) the distance between the two chromatids in one chromosome; and (3) the distance between the two poles of the cell. These three variables are illustrated schematically in the graph.

The illustrations used in Fig. 6.15 are tremendously helpful for readers to recognize what data are being measured. But for readers who are not already familiar with cell structure and division, to try to correlate the changing data lines with various cellular components and distances and, moreover, to envision a big picture of how these values correlate at different stages remain difficult. In this case, pictographs, by forgoing the formal appearance of Cartesian graphs and foregrounding the visual context of quantitative data, can convey those data in richer and more accessible ways, as Fig. 6.16 shows.

In Fig. 6.16, readers can directly observe the changes that happen during cell division; can measure, using the built-in grid, the extent of those changes; and can compare how one change impacts or relates to another at different times. For example, when, 15 min into the process, the two chromatids in a chromosome separate and move to their respective poles, the distance between the chromatids notably increases, while

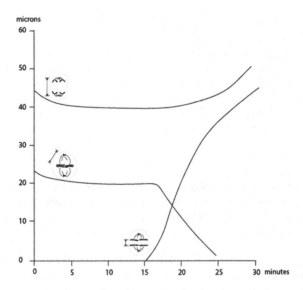

Fig. 6.15 Conventional graph of structural changes during cell division. Recreated graph represents only general data trends. Created based on Mazia (1961, p. 115)

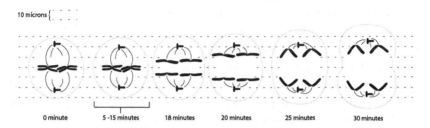

Fig. 6.16 Pictograph of structural changes during cell division

that between a chromatid and its pole notably decreases; meanwhile, the distance between the two poles increases too, but only slightly. These trends, which are buried in the Cartesian coordinate system (though are definitely present and probably obvious for select readers), now become almost commonsensical.

While I believe pictographs are a missed opportunity, I am not suggesting that we use them for all quantitative evidence in popular science communication. Conventional graphs have their advantage in presenting precise data with minimalist designs. In addition, depending on context, some data, trends/patterns, and underlying concepts may well resist intelligent and engaging pictographic transformation. What I do argue is that we be willing to reconsider our bias against pictographs and be willing to resist a default resort to Cartesian graphs.

Conclusion

Designed to demonstrate quantitative trends and patterns, Cartesian graphs are an inevitable genre in the communication of big data science. In popular communication, they provide readers with opportunities to participate in "making science," to examine as well as evaluate how conclusions are formed based on data evidence. On the other hand, despite being commonly seen in everyday life, graphs *can* be difficult to access when they demand, unreasonably so, that publics be familiar with experiment procedures, formal jargon, and unspoken conventions. For these reasons, graphing for public readers cannot simply be a mathematical maneuver but, to borrow the words of the Isotype team, has to be a transformative process of telling meaningful, relatable, and engaging stories about data. Doing so may mean choosing to graph concrete data that are already relatable to public readers, supplementing quantitative information with visual entry points, elevating technical details to higher-level data relationships, or using alternative methods that disregard the Cartesian formality. Some of these heuristics may well add to a graph's visual and emotional appeal, but equally, if not more importantly, they afford graphs cognitive interest, semantic relevance, and pragmatic engagement.

There is, of course, another alternative: not to graph at all. While we celebrate graphs' ability to present data, more data is not necessary superior, especially when we consider that prose is more adept at explaining abstract information. Figure 6.5 that entailed an extensive walk-through offers one such occasion. In that case, the obscure graph may be more easily replaced by verbal explanations such as "Two kinds of DNA, one lighter and one heavier, are found in the *Chlamydomonas* whole cell and chloroplast. The lighter DNA accounts for a small amount of whole cell DNA but a significant amount of chloroplast DNA, which suggests that this DNA is uniquely situated in the chloroplast." If desired, details about centrifugation, UV absorption, and data measurement can also be verbally included.

As this chapter shows, considerable research on Cartesian graphs exists in the areas of data visualization, information design, cognitive science, and risk communication, but the focuses of these studies do not necessarily parallel the concerns of popular science communication. As a whole, these works valorize "proper" Cartesian graphs and prioritize, implicitly or otherwise, specialists' information needs. Or, they focus on the context of clinical practice and thus take for granted the need to "properly educate" readers about graphs. For scientists and popular science communicators to entertain alternative graphing approaches and priorities, we need studies that explicitly examine public readers' attitudes, preferences, and information needs. I hope this chapter contributed to that goal and broached topics and heuristics for future research.

Notes

1. Figure 6.2 lacks a vertical axis so readers cannot know the exact number of plants that bear a certain number of flowerets.
2. CCR5 gene encodes a protein that HIV binds to in order to infect host cells.
3. Kosslyn (1989) acknowledged that, occasionally, a patterned background such as a photograph may add to the affective interpretation of a graph, though it may also interfere with the data display.
4. This changes when we consider color-deficient viewers (see Chap. 3).
5. Found in red blood cells, hemoglobin molecules transport oxygen from respiratory organs to body tissues.
6. Figure 6.8 adaptation may well have been guided by this same intention and resulted in data reduction.
7. The original study examined many other genes.
8. Isotype itself may be attributed to late 19th century prototypes (see Neurath and Kinross 2009).
9. As Miller et al. (2016) showed, emoji can be interpreted differently by difference people. When using face icons for pictographs, choosing icons that show basic emotions and supplementing them with textual annotations help to reduce confusion.
10. 23andMe launched its direct-to-consumer genetic testing in 2007; in 2013, it was required by the U.S. Food and Drug Administration (FDA) to discontinue this service in the U.S. market. But in February 2015, FDA authorized the company to market a carrier test, which determines whether healthy individuals carry a genetic variant that could cause their offspring to inherit Bloom Syndrome (U.S. Food and Drug Administration 2015).
11. deCODEme launched its personal genome analysis service in 2007, providing individual consumers with information about their disease risks. In 2012, when the company was sold to Amgen, this service was stopped.

References

Amare, N., & Manning, A. (2013). *A unified theory of information design: Visuals, text & ethics.* Amityville, NY: Baywood Publishing.

Ancker, J. S., Senathirajah, Y., Kukafka, R., & Starren, J. B. (2006). Design features of graphs in health risk communication: A systematic review. *Journal of the American Medical Informatics Association, 13*(6), 608–618.

Ayala, F. J. (1974). Biological evolution: Natural selection or random walk. *American Scientist, 62*(6), 692–701.

Ayala, F. J., & Anderson, W. W. (1973). Evidence of natural selection in molecular evolution. *Nature New Biology, 241*(113), 274–276. doi:10.1038/newbio241274a0.

Bateman, S., Mandryk, R. M., Gutwin, C., Genest, A., McDine, D., & Brooks, C. (2010). Useful junk?: The effects of visual embellishment on comprehension and memorability of charts. In *Human Factors in Computing Systems Proceedings* (pp. 2573–2582). New York: ACM.

Bauer, M. W., & Gaskell, G. (2008). Social representations theory: A progressive research programme for social psychology. *Journal for the Theory of Social Behaviour, 38*(4), 335–353. doi:10.1111/j.1468-5914.2008.00374.x.

Bauer, W. R., Crick, F. H. C., & White, J. H. (1980). Supercoiled DNA. *Scientific American, 243*(1), 118–133.

Berg, C. A., & Phillips, D. G. (1994). An investigation of the relationship between logical thinking structures and the ability to construct and interpret line graphs. *Journal of Research in Science Teaching, 31*(4), 323–344. doi:10.1002/tea.3660310404.

Borel, E. (1914, December 26). "Heads and tails" and heredity. *Scientific American,* 403–405.

Borgo, R., Abdul-Rahman, A., Mohamed, F., Grant, P. W., Reppa, I., Floridi, L., & Chen, M. (2012). An empirical study on using visual embellishments in visualization. *IEEE Transactions on Visualization and Computer Graphics, 18*(12), 2759–2768.

Brasseur, L. E. (2003). *Visualizing technical information: A cultural critique.* Amityville, NY: Baywood.

Brown, K. (2001). Seeds of concern. *Scientific American,* 52–57.

Burke, C. (2009). Isotype: Representing social facts pictorially. *Information Design Journal, 17*(3), 211–223.

Cates, C. (2008). *Visual Rx version 3.* Retrieved September 19, 2015, from http://www.nntonline.net/visualrx/.

Commoner, B. (1964). DNA and the chemistry of inheritance. *American Scientist, 52*(3), 365–388.

Condit, C. M. (1999). *The meanings of the gene: Public debates about human heredity.* Madison: University of Wisconsin Press.

Conklin, E. (1914, October). Phenomena of inheritance. *The popular science monthly,* 313–337.

DelMas, R., & Liu, Y. (2005). Exploring students' conceptions of the standard deviation. *Statistics Education Research Journal, 4*(1), 55–82.

Dyson, F. (2015). *Dreams of earth and sky.* New York: New York Review Book.

Edwards, A., Elwyn, G., & Mulley, A. (2002). Explaining risks: Turning numerical data into meaningful pictures. *BMJ, 324*(7341), 827–830.

Elmore, J. G., & Gigerenzer, G. (2005). Benign breast disease–the risks of communicating risk. *The New England Journal of Medicine, 353*(3), 297–299.

Fagerlin, A., Wang, C., & Ubel, P. A. (2005). Reducing the influence of anecdotal reasoning on people's health care decisions: Is a picture worth a thousand statistics? *Medical Decision Making: An International Journal of the Society for Medical Decision Making, 25*(4), 398–405.

Galesic, M., Garcia-Retamero, R., & Gigerenzer, G. (2009). Using icon arrays to communicate medical risks: Overcoming low numeracy. *Health Psychology: Official Journal of the Division of Health Psychology, American Psychological Association, 28*(2), 210–216. doi:10.1037/a0014474.

Garcia-Retamero, R., Galesic, M., & Gigerenzer, G. (2010). Do icon arrays help reduce denominator neglect? *Medical Decision Making, 30*(6), 672–684.

Gesellschaft und Wirtschaft: Bildstatistisches Elementarwerk. (1930). Leipzig: Bibliographisches Institut AG.

Gibson, J. M., Rowe, A., Stone, E. R., & de Bruin, W. B. (2013). Communicating quantitative information about unexploded ordnance risks to the public. *Environmental Science and Technology, 47*(9), 4004–4013. doi:10.1021/es305254j.

Goodenough, U. W., & Levine, R. P. (1970). The genetic activity of mitochondria and chloroplasts. *Scientific American,* 22–29.

Gorman, C., & Maron, D. (2014, April). The RNA revolution. *Scientific American,* 52–59.

Gorr, T., & Kleinschmidt, T. (1993, January–February). Evolutionary relationships of the coelacanth. *American Scientist, 81*(1), 72–82.

Gowen, J. (1935, September). Genetic constitution as a factor in disease. *Sigma Xi Quarterly, 23*(3), 103–117.

Griffiths, A., Miller, J., Suzuki, D., Lewontin, R., & Gelbart, W. (2000). *An introduction to genetic analysis* (7th ed.). New York: W. H. Freeman.

Gross, A. G., & Harmon, J. E. (2013). *Science from sight to insight: How scientists illustrate meaning.* Chicago, IL: University of Chicago Press.

Haard, J., Slater, M. D., & Long, M. (2004). Scientese and ambiguous citations in the selling of unproven medical treatments. *Health Communication, 16*(4), 411–426. doi:10.1207/s15327027hc1604_2.

Haroz, S., Kosara, R., & Franconeri, S. (2015). ISOTYPE visualization: Working memory, performance, and engagement with pictographs. *Human Factors in Computing Systems Proceedings* (pp. 1191–1200). New York: ACM.

Jacques, B. (1983). *Semiology of graphics.* (W. J. Berg, Trans.). Madison: University of Wisconsin Press.

Jordan, V. C. (1998). Designer estrogens. *Scientific American*, 60–67.

Kosslyn, S. M. (1989). Understanding charts and graphs. *Applied Cognitive Psychology, 3*(3), 185–225. doi:10.1002/acp.2350030302.

Kosslyn, S. M. (2006). *Graph design for the eye and mind.* New York: Oxford University Press.

Landauer, W. (1940). The nature of disproportionate dwarfism, with special reference to fowl. *Sigma Xi Quarterly*, 171–180.

Leinhardt, G., & Others, A. (1990). Functions, graphs, and graphing: Tasks, learning, and teaching. *Review of Educational Research, 60*(1), 1–64.

Lloyd, F., & Reyna, V. (2001). A web exercise in evidence-based medicine using cognitive theory. *Journal of General Internal Medicine, 16*(2), 94–99.

Mattick, J. (2004). The hidden genetic program of complex organisms. *Scientific American*, 60–67.

Mazia, D. (1961). How cells divide. *Scientific American,* 100–120.

Mendel, J. G. (1866). Versuche über Pflanzen-Hybriden. *Verhandlungen des naturforschenden Vereines in Brünn, 4,* 3–47.

Miller, H., Thebault-Spieker, J., Chang, S., Johnson, I, Terveen, L., & Hecht, B. (2016). "Blissfully happy" or "ready to fight": Varying interpretations of emoji. *Proceedings of the 10th International AAAI Conference on Web and Social Media* (pp. 259–268). Palo Alto, California: AAAI Press.

Morse, G. (1984). Genetic engineering and the jumping gene. *Science News, 125*(17), 264–265, 268.

Nestler, E. J. (2011). Hidden switches in the mind. *Scientific American*, 76–83.

Neurath, M., & Kinross, R. (2009). *The transformer: Principles of making Isotype charts.* London, UK: Hyphen Press.

O'Brien, S. J., & Dean, M. (1997). In search of AIDS-resistance genes. *Scientific American*, 44–51.

Ostrander, E. A. (2007, September–October). Genetics and the shape of dogs. *American Scientist, 95*(5), 406–413.

Paling, J. (2003). Strategies to help patients understand risks. *BMJ (Clinical Research Ed.), 327*(7417), 745–748.

Perera, F. P. (1996, May). Uncovering new clues to cancer risk. *Scientific American,* 54–62.

Pozzer, L., & Roth, W. (2003). Prevalence, function, and structure of photographs in high school biology textbooks. *Journal of Research in Science Teaching, 40*(10), 1089–1114. doi:10.1002/tea.10122.

Randy, K. (2013). *Cool infographics: Effective communication with data visualization and design.* Hoboken, NJ: Wiley.

Roth, W., & Mcginn, M. K. (1997). Graphing: Cognitive ability or practice? *Science Education, 81*(1), 91–106. doi:10.1002/(SICI)1098-237X(199701)81:1<91:AID-SCE5>3.0.CO;2-X.

Saey, T. H. (2010, December 18). Genetic dark matter: Searching for new sources to explain human variation. *Science News, 178*(13), 18–21.

Saey, T. H. (2012, March 10). Centenarians distinguished by DNA signatures. *Science News, 181*(5), 20.

Sager, R. (1964). Genes outside the chromosomes. *Scientific American,* 70–79.

Sebastiani, P., Solovieff, N., DeWan, A. T., Walsh, K. M., Puca, A., Hartley, S. W., … Perls, T. T. (2012). Genetic signatures of exceptional longevity in humans. *PLoS ONE, 7*(1), e29848. doi:10.1371/journal.pone.0029848.

Smerecnik, C. M. R., Mesters, I., Kessels, L. T. E., Ruiter, R. A. C., De Vries, N. K., & De Vries, H. (2010). Understanding the positive effects of graphical risk information on comprehension: Measuring attention directed to written, tabular, and graphical risk information. *Risk Analysis, 30*(9), 1387–1398. doi:10.1111/j.1539-6924.2010.01435.x.

Sudmant, P. H., Kitzman, J. O., Antonacci, F., Alkan, C., Malig, M., Tsalenko, A., … Eichler, E. E. (2010). Diversity of human copy number variation and multicopy genes. *Science, 330*(6004), 641–646.

Tal, A., & Wansink, B. (2014). Blinded with science: Trivial graphs and formulas increase ad persuasiveness and belief in product efficacy. *Public Understanding of Science,* 1–9, doi:10.1177/0963662514549688.

Tufte, E. R. (2001). *The visual display of quantitative information.* Cheshire, CT: Graphics Press.

U.S. Food and Drug Administration. (2015, February 19). *FDA permits marketing of first direct-to-consumer genetic carrier test for Bloom syndrome.* Retrieved October 8, 2015, from http://www.fda.gov/NewsEvents/Newsroom/Press Announcements/ucm435003.htm.

Waters, E., Weinstein, N., Colditz, G., & Emmons, K. (2006). Formats for improving risk communication in medical tradeoff decisions. *Journal of Health Communication, 11*(2), 167–182. doi:10.1080/10810730500526695.

Weisberg, D. S., Keil, F. C., Goodstein, J., Rawson, E., & Gray, J. R. (2008). The seductive allure of neuroscience explanations. *Journal of Cognitive Neuroscience, 20*(3), 470–477.

Wong, K. (2014, September). The 1 percent difference. *Scientific American,* 100.

7

The Structural View: 2D Realities and 3D Possibilities

The 1950s, in genetics years, marked the beginning of the DNA era. Various studies, notably Hershey–Chase's (1952) experiment on phage virus, proved that DNA molecules are the functional element that carries genetic information. Shortly after, in 1953, James Watson and Francis Crick deduced DNA's double-helix structure, which revealed the mechanism of how genetic information is stored and duplicated. Widespread interest and research on the molecular structure of gene products followed. This is because molecular structures, as in the case of DNA and especially in the case of proteins, are closely associated with biological functions. Being able to solve the structure of gene products thus paves the way for targeted drug development and genetic medicine. This endeavor, in the post-Human Genome Project era, culminated in the subdiscipline of structural genomics, which aims to solve the 3D structures of all gene products in a given genome. Given the availability of enormous sequence data (see Chap. 5), researchers try to compare the sequences of different proteins to predict structures based on previously solved ones.

Communicating these research developments to public audiences is, itself, a decades-long endeavor. Given that this line of research focuses

© The Author(s) 2017
H. Yu, *Communicating Genetics*,
DOI 10.1057/978-1-137-58779-4_7

on structural arrangements, visual representations play a significant role in that communication. However, molecular structures, which are often quite intricate and different from every day artifacts, are not readily familiar or easily relatable to public readers. The presentation of these structures is also bound by the limits of the 2D medium and later, the reality and availability of computer-based visualization programs. Examining the multiple and evolving ways molecular structures are visualized in the popular communication of genetics, this chapter raises questions about this fundamental but understudied area of popular science visualization and suggests areas for future research.

2D Structural Formulas

Structural formulas are 2D graphics that use established chemical symbols to represent the atoms that make up a molecule as well as the ways those atoms are bonded. They differ from a text-based molecular formula, which is a linear description of all the atoms in a molecule without indications of spatial arrangements. For example, the molecular formula of water is simply H_2O, but a structural formula would also demonstrate how the two hydrogen atoms (H) and the oxygen (O) in a water molecule are positioned in relation to each other, as in H:O:H.

Though the history of structural formulas is hard to pinpoint, representations similar to what we see today had started to appear in early nineteenth century (Bradshaw 2001). They became an essential tool in the late nineteenth century with the development of the structural theory of chemistry. Indeed, late nineteenth century also marked structural formulas' foray into popular attention, thanks to German chemist August Kekulé's (1872) famous dream of a snake seizing its own tail, which inspired Kekulé to discover the ring-shaped structure of benzene. Today, structural formulas are still commonly used in professional journals and textbooks and provide a convenient way to visualize the formation of a substance as well as to explain or predict its function (Cooper et al. 2010; O'Donoghue et al. 2010).

In the popular communication of genetics, structural formulas were most heavily used around the 1950s–1960s, depicting the structure of genetic elements, their structural interactions with each other, and their interactions with other chemical compounds, whether harmful or therapeutic. These visual displays were still used during the 1970s–1980s and the following decades, but toward the last two decades of the twentieth century, when computer-generated models appeared (more about this later), structural formulas became rare, except for when depicting the single most famous genetic element, the DNA molecule, or compounds with a similar structure.

Of course, structural formulas are not one single type of visual representation. There are, in fact, a number of formula systems. The H:Ö:H representation of water molecules, for instance, is an example of the Lewis dot formula. Established by Gilbert Newton Lewis, the system uses standard symbols to represent atoms (O for oxygen, H for hydrogen, etc.); it then adds small dots around each atom to represent electrons (known as valence electrons) that are capable of forming bonds with other atoms (Lewis 1916). Water (H_2O) appears as H:Ö:H because oxygen (O) has six valence electrons and each hydrogen (H) has one valence electron. In addition to this graphing system, there are also the skeletal formula, the sawhorse projection, the Newman projection, the chair conformation, etc. Not all of these systems, however, are commonly used in the popular communication of genetics. The two systems that are, possibly due to their relatively simpler visual appearance, are the line-bond formula and the skeletal formula.

Line-Bond Formula

An easy way to describe the line-bond formula is to see it as a simplification of the Lewis dot formula mentioned above. That is, every two electrons/dots that form a bond between two atoms is simplified into a short line. H_2O or H:Ö:H thus becomes H–Ö–H. For additional simplification, the four non-bonding electrons around the oxygen can be left out, which turns the display for water into H–O–H. A reader deduces the omitted electrons by knowing how many valence electrons surround an oxygen or by following the octet rule, which states that common atoms such as carbon and oxygen would be surrounded by eight electrons.[1]

A line-bond formula may also contain double lines, which indicate a stronger, double bond that involves four electrons.

Figure 7.1 is an example of the line-bond formula, depicting, in this case, the structure of the DNA molecule. Placed within the two "wavy lanes" at the top and bottom of the figure are DNA's two backbone strands, each consisting of phosphate groups and sugars. Connecting the two strands are four DNA bases: thymine, cytosine, adenine, and guanine. Of these four bases, thymine always bonds with adenine, and cytosine always bonds with guanine.

As Fig. 7.1 shows, in the line-bond formula, atoms can be visualized as "balls," a common and intuitive, if inaccurate, metaphorical approach to explaining particles (Baake 2003). The bonds between atoms appear as connecting "lines," which are also metaphorically intuitive and easy to imagine. Electrons are removed from the depiction, which reduces

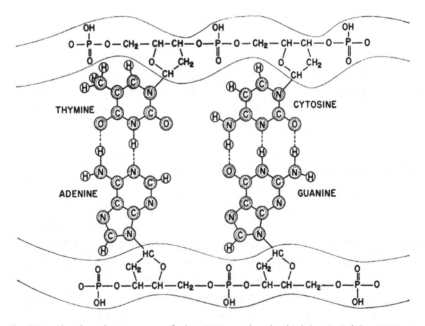

Fig. 7.1 Line-bond structure of the DNA molecule (Calvin & Calvin 1964, p. 173). Reprinted by permission of *American Scientist*, magazine of Sigma Xi, The Scientific Research Society

visual complexity and helps to focus readers' attention on atoms and their spatial relationships.

Skeletal Formula

Though the line-bond formula simplified the Lewis dot structure, it can still, as shown in Fig. 7.1, be visually complex when used to depict genetic research. This is because DNA and gene products, notably proteins, are macromolecules that contain large numbers of atoms and bonds—and hence, require elaborate visualization. This is where the skeletal formula comes in. In a skeletal formula, carbons (C), the most common and essential atom in organic compounds, are omitted. Unless otherwise specified, the end of a line or the junction of two lines is taken to be a carbon. The simple display of ⋀ therefore, represents three carbons.

In addition, hydrogen (H), the atom most likely to bond with carbon, is omitted when it does bond with carbon. To deduce omitted carbon–hydrogen bonds, one needs to know that a carbon has four valence electrons to form four bonds. Any unspecified bonds within this limit of four are thus omitted carbon–hydrogen bonds. In ⋀ the middle carbon would have two omitted hydrogen bonds because this carbon already forms two bonds with the two end carbons. Likewise, each end carbon would have three hydrogen bonds because it only has one bond with the middle carbon. Thus, the simple display of ⋀ is the equivalent of H-C-C-C-H. For further simplification, when hydrogen atoms bond with non-carbon atoms, the lines between the atoms may be omitted. Thus, a O–H bond becomes OH.

Given this system, the DNA structure in Fig. 7.1 becomes that shown in Fig. 7.2. This visual simplification helps to emphasize the structural pattern of the molecule. For instance, it is now more obvious that each DNA backbone strand consists of two alternating components (i.e., a phosphate group and a 5-carbon sugar[1]). The structural similarities as well as differences

---

[1]A minus sign indicates that an atom is negatively charged.

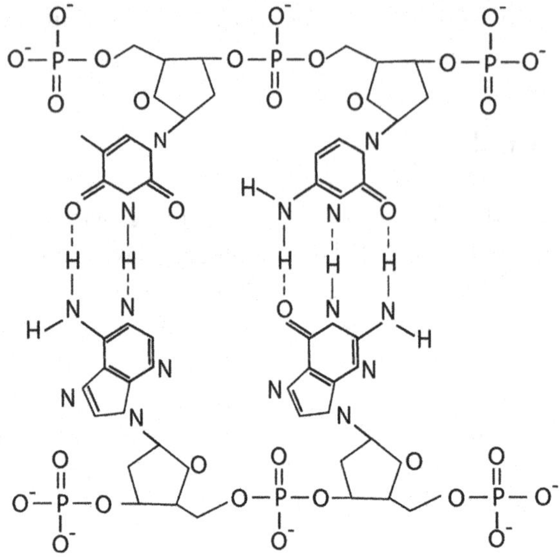

Fig. 7.2 Skeletal structure of the DNA molecule

among the four DNA bases are also more apparent. Given these advantages, the skeletal formula is often the formula of choice in the popular communication of genetics (see e.g., Bearn 1956; Miller 1977).

(Not) Making Sense of Structural Formulas

As the above discussion makes clear, structural formulas, regardless of which system they use, fall under the category of *symbol* in Peirce's (1894) sign system; that is, their meaning depends on established representation schemes, disciplinary conventions, and prior knowledge. To be able to derive their meaning, then, readers must be aware of those conventions and knowledge as well as their complex interplay (Cooper et al. 2010). Such conventions and knowledge are not necessarily more difficult than those involved in other elaborate symbolic systems (such as our language system), but they are not something all public readers, especially readers who have no formal training in chemistry, can be expected to be familiar with. Indeed, the literature suggests that even university chemistry students and faculty, despite their levels of formal training, do not have enough understanding of structural formulas (Cooper et al. 2010). Compounding these barriers is the number of elements often involved in structural formulas—even the skeletal structure, as shown in Fig. 7.2, still must utilize a fairly large number of syntactic elements, which deter visual processing. And paradoxically, efforts to reduce surface-level visual complexity arguably make the images less accessible, as viewers must be equipped with more prior knowledge in order to deduce the omitted visual elements.

For these reasons, structural formulas can come across, to borrow Amare and Manning's (2013) words, as a "complex block of lines…and shapes" that makes a public reader "*feel* that the visual is informative" without actually being so (p. 111). Accompanying this inaccessibility is a sense of affective distance (see Christiansen 2013), a feeling that such images and their meanings are simply not something that would appeal to public readers. Given these, it is not surprising that structural formulas are disappearing in contemporary popular science magazines.[2] If earlier publications were guided by an enthusiasm to share with publics all the latest genetic research and a conviction that readers will automatically find those findings worthy of careful study, today's scientists and science communicators seem to have become more aware of public readers' own cognitive and affective needs.

But to stop at this level of analysis seems to be simplifying the matter—not to mention resigning a fundamental point of public science communication: to find ways to communicate otherwise complex and unfamiliar information in accessible and engaging manners. If structural formulas have sustained centuries of refinement and been found useful by scientists to construct and disseminate molecular information, *can* these images be of value in popular communication? Moreover, if such images offer a convenient way to graph the structure of gene products and if structural information is essential to understanding gene functions, *should* we find a place for such images in popular communication?

One way to explore these questions is to examine structural formulas as they were used in their "heyday." Figure 7.3 is one example. Using the skeletal formula,[3] Fig. 7.3 attempts to explain the working mechanism of AZT, the first drug found to be able to curb AIDS and the first U.S. government-approved AIDS treatment. According to the image's

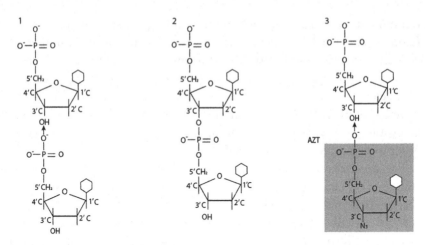

Fig. 7.3 Skeletal structure of how AZT curbs HIV viral DNA. Created based on Gallo (1986, p. 54)

original caption, the HIV viral DNA (or any DNA for that matter), shown in part 1 of the formula, consists of subunits called nucleotides; each nucleotide includes a sugar molecule, which has five carbons. The third sugar carbon has a hydroxyl group (OH), which helps to bond two nucleotide subunits and extend the DNA chain, as shown in part 2. Because AZT is structurally similar to a nucleotide but lacks the OH group, it can be incorporated into the viral DNA but will then terminate the DNA and inactivate the virus, as shown in part 3.

Figure 7.3 is a prime example of how structural formulas *can* be of value in popular science communication: They put into concrete terms the biochemical nature of genetic research and help to demystify the discipline; they also demonstrate how molecular structure correlates with biological functions and how structural research has important implications for medical research. That said, trying to follow Fig. 7.3 and map the previous caption onto the image is not necessarily a straightforward process. Not only are there a large number of individual letters, lines, and shapes, but these visual elements must also be collectively processed in relation to each other for the image to make sense.

For example, readers need to first identify which elements constitute the nucleotide, how multiple nucleotides combine, via what mechanism, and then how that combination is disrupted, via what mechanism. When materials contain many elements, especially highly interactive elements that cannot be processed in isolation, such materials are said to pose high cognitive load (Sweller 1994; Cook 2006). And because an individual's cognitive capacity is limited and can only actively process a small amount of information at a given time (Miller 1994; Sweller 1994), materials such as Fig. 7.3 are inherently difficult to follow.

Readers who are experienced in a subject matter do not feel the effect of complex element interactivity because they automatically incorporate elements into schemas (Sweller 1994). Each schema "can hold a large amount of information, [but] it is processed as a single unit in working memory" and thus reduces the burden on cognitive load (Cook 2006, p. 1076). In the case of structural formulas, the various letters, lines, and shapes do not appear as discrete elements but readily suggest themselves as schemas. For example, for a geneticist, a quick glance at Fig. 7.3 will reveal such commonly referenced objects as nucleotide units and 5-carbon sugars. If this processing mechanism can be reproduced for public readers, that would help bypass the inherent, high cognitive load posed by structure formulas and make these images more accessible.

To do so means ostensibly presenting the discrete elements in a structural formula as patterned units so readers who do not have ready-made schemas will see them as such. Reading structural formulas would then become a matter of tracking and comparing visual patterns, a much easier way to process complex information (Ware 2012). Indeed, Fig. 7.3 started this strategy by "boxing" AZT into a single unit but did not apply it to other components, possibly because of a (groundless) assumption that readers can easily process all the other, "familiar" structural components. A more thorough use of the schema strategy leads to a possible revision in Fig. 7.4, which displays the same structural formulas but uses boxes and shading to create artificial boundaries around key structural units: namely, the nucleotide, the 5-carbon sugar, the OH group, the AZT, and the N_3 group in AZT. Although there are multiple visual elements within each box and shaded area, these elements are

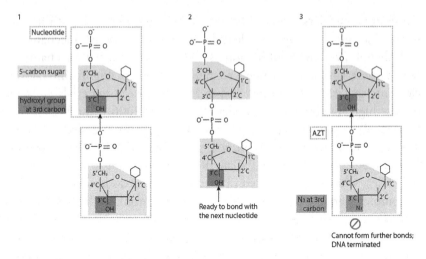

Fig. 7.4 Schema-based structural representation of how AZT curbs HIV viral DNA (color figure online)

effectively seen as a single schema, an effect caused by the Gestalt principle of enclosure, which states that our eyes see physically enclosed visual elements as a coherent single unit (Gross 2007).

As a result, in Fig. 7.4 part 1, we see, more clearly than before, two nucleotides, each having a 5-carbon sugar with an OH group at the third carbon position. Part 2 shows the two nucleotides bonded at the OH group via what is known as a phosphodiester bond; part 2 then highlights the OH group of the second nucleotide, which is ready for further bonding. In part 3, the structural similarity between the AZT and nucleotide becomes obvious, as is their difference: instead of an OH group, ATZ has an N_3. This structural difference explains how the drug stops the further bonding of nucleotides and disrupts the HIV viral DNA.

The visual schemas in Figs. 7.3 and 7.4 are relatively easy to determine, given that structures such as nucleotides and 5-carbon sugars are commonly referenced, standard subunits in DNA. With other organic compounds, the determination of schemas may be less obvious. But this does not prevent an author from creating schemas that, albeit "non-standard," make sense in a given communication context. These schemas may be created based on formulas' visual similarity, which is a basic Gestalt principle that aids pattern recognition (Johnson 2010). Or they may be created based on formulas' *lack* of similarity, in other words, their contrast, which is another Gestalt principle that allows pattern recognition and comparison (Gross 2007).

Consider Fig. 7.5, which depicts the three routes via which the human body creates guanylic acid, a substance necessary to form DNA and RNA. The de novo route (left to right) converts inosinic acid into guanylic acid. The other two routes, known as reutilization routes, both require an enzyme called HGPRT. In reutilization route 1 (top left to right), the enzyme helps to convert hypoxanthine into guanylic acid.

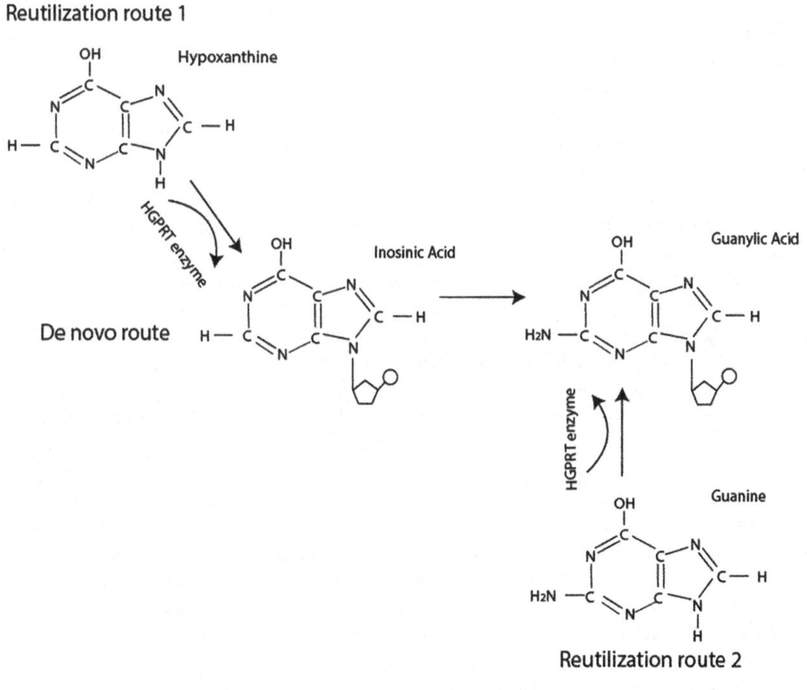

Fig. 7.5 Three routes to create guanylic acid. Created based on Friedmann (1971, p. 40)

In reutilization route 2 (bottom right up), the enzyme helps to convert guanine into guanylic acid. Because a certain region of the brain depends on the enzyme-mediated routes for guanylic acid, individuals with a defective gene for the HGPRT enzyme would exhibit impaired brain function, resulting in Lesch–Nyhan syndrome (Friedmann 1971).

Again, for readers unfamiliar with the structures of the various compounds in question, Fig. 7.5 does not readily reveal visual schemas, which makes it difficult to discern what happened in each route, how the three routes compare, and what the role of the HGPRT enzyme is. Although in this case, there are no necessarily standard chemical subunits, visual schemas can be created by highlighting structural similarities as well as differences. Figure 7.6 demonstrates a possible way of doing so.

The complex chemical processes now more clearly reveal their nature. For instance, in the de novo approach, the conversion from inosinic

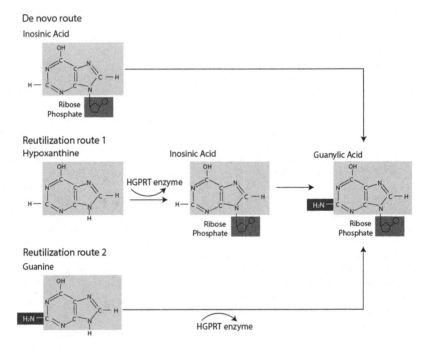

Fig. 7.6 Schema-based structural representation of the three routes to create guanylic acid (color figure online)

acid to guanylic acid is due to the change of one chemical unit (from –H to –H$_2$N). This same process happens at the second step of reutilization route 1. Or, in reutilization route 2, the addition of a ribose phosphate marks the difference between guanine and guanylic acid, and a comparable process happens at the first step of reutilization route 1. It is useful to note that the point of highlighting these mechanisms is not so that public readers will know or remember how, for example, guanylic acid is chemically structured. To the contrary, the point of organizing detailed structural information into patterns is precisely so that readers can be unaware of those details but still find the formula cognitively interesting and relevant. In this case, the display is relevant because it allows readers to recognize why genetic deficiency leads to neurological diseases and, by extension, how genetic research and medicine may present possibilities for intervention. By conveying such messages in concrete terms, structural formulas allow scientists to articulate the larger social impact of their work.

Though structural formulas continue to be a common and important visual display in formal scientific discourses and basic genetic research, they are no longer common in today's popular communication of genetics. This trend may well have to do, at least partly, with scientists' and science communicators' desire to make genetic research accessible to public readers who do not have formal training in chemistry. While there is no denying the merit of such an intention, it is also pertinent to consider whether doing so may, at times, take away from the valuable information in popular communication and further demarcate the division between "real" and popular science. Though we cannot change the formally codified nature of structural formulas and their inherent cognitive load, we may, as I argued above, attempt to present them in ways that facilitate public conversation.

3D Molecular Models

The difference between 2D structural formulas as we discussed above and 3D molecular models can be fuzzy—especially when we consider that in print and static digital media, 3D objects appear in two dimensions. Indeed, for some scholars, structural formulas *are* the earliest form of molecular models (Schlecht 1997). More generally, however, the term "molecular model" refers to physically assembled and, later, computer-simulated artifacts that display the structures of molecules in three dimensions. German chemist August Wilhelm von Hofmann is often credited to be the first person to use physical molecular models during his 1865 lecture before the Royal Society of Great Britain (Schlecht 1997). Following Hofmann, a series of important work on the 3D nature of molecules was completed in the nineteenth-to-mid-twentieth century, allowing 3D displays to be appreciated by the chemistry community and deemed crucial to their research activities (Schlecht 1997).

What made molecular models a well-known public artifact, however, was not chemistry but genetics. In the early 1950s, in the race to uncover the structure of DNA, researchers around the world explored different methods. Among them was physical modeling, the method of choice by the ultimate winner of the race, James Watson and Francis Crick, as well as one of their greatest rivals, Linus Pauling (Pray 2008). Using cardboard cutouts, metal plates, and rods to represent "individual chemical components of the four bases and other nucleotide subunits, Watson and Crick shifted molecules around on their desktops, as though putting together a puzzle" (Pray 2008). With trials and errors—and crucial insights gained from Rosalind Franklin's DNA X-ray crystallography and Erwin Chargaff's work on DNA bases (see Chaps. 2, 5)—Watson and Crick eventually deduced DNA's double-helix structure. A reconstruction of their model, using many of the original metal plates, is now prominently on display in the London Science Museum. In print and online media, popular communication of genetics does not allow such monumental exhibits, but what they can and do offer is no less intriguing.

Perspective Drawings

In the popular communication of genetics, the earlier depictions of 3D models took the form of perspective drawings (see e.g., "double helix" 1973; Miller 1977). Figure 7.7, for example, again shows the familiar

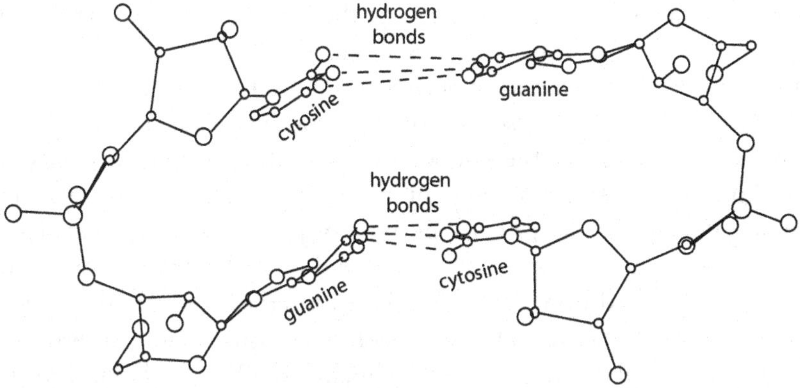

Fig. 7.7   Perspective drawing of DNA molecule

DNA molecule with its two sugar-phosphate strands; the strands are connected by DNA base pairs via hydrogen bonds. Both of the DNA base pairs shown in the image are cytosine and guanine pairs, though another kind of pair, that between thymine and adenine, is also possible.

Rather than standard chemical symbols, Fig. 7.7 relies more heavily on common visual elements such as "balls" (for atoms), "sticks" (for bonds), and spatial positions to depict its molecular structure. For that reason, it is more visually concrete and requires less background knowledge or awareness of formal conventions from a viewer. At the same time, because of molecules' complex composition and the constraints of perspective drawings (e.g., objects will diminish with distance, and certain lines must converge), such depictions are not necessarily clear to view. In the case of Fig. 7.7, while the DNA backbone structures are more or less visible from a frontal view, most of its base structures are not.

Computer-Simulated Models

Molecular visualization software, with its ability to generate optically "true" 3D images, addressed some of these perceptual issues. Computer programs capable of simulating molecular models became a technical reality toward the last two decades of the twentieth century (Schlecht 1997). Around the same time was when 3D models began to replace structural formulas in the popular communication of genetics. These 3D models, as in the case of structural formulas, are not one single type of visual display. Their appearance, to a large extent, depends on the functionalities and settings of the software used to create them. Some software has the primary purpose of presenting molecular structures, while others allow users to build and modify a model; some programs are accessed from command-driven interfaces, while others allow mouse-driven access (details about molecular visualization software see Tate [2003] and Bottomley and Helmerhorst [2009]). From a visual setting perspective, most programs have the ability to display molecules in a range of "styles" and to allow user interactions such as rotation and zooming (Bottomley and Helmerhorst 2009).

Of these settings, the "style" assumed by a model is the aspect immediately visible and thus significant to a viewer. It is also one of the most important choices a visual creator can make, because different styles help to emphasize different structural details. Common molecular modeling styles include the wireframe/stick style, which uses lines to represent bonds and the terminals of lines to represent atoms; the ball-and-stick style, which uses balls to represent atoms and sticks to represent bonds; the space-filling style (Fig. 7.8 top), which uses interconnected spheres to represent atoms; the surface style, which uses smooth surfaces or meshes to represent the overall shape of a molecule; and the ribbon style (Fig. 7.8 bottom), which uses ribbons, arrows, and ropes to represent different parts of a molecule (Bottomley and Helmerhorst 2009; Gasgall 2008). All of these styles appear in the contemporary popular communication of genetics.

Compared with perspective drawings, images such as those in Fig. 7.8 are more concrete, precise, and clearly detailed; what one sees seems the exact molecular reality that is "out there." Moreover, syntactic details such as color, shading, lighting, texture, and transparency can be easily added to distinguish different structural elements and create a realistic 3D appearance (Bottomley and Helmerhorst 2009). Figure 7.8 top, for example, uses color-coded spheres to distinguish the different atoms that form DNA. Lightness contrasts are used to highlight parts of the

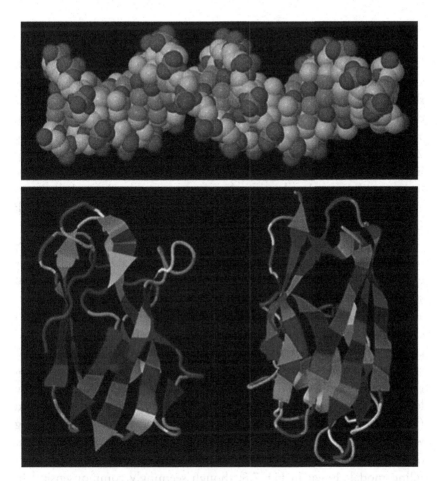

Fig. 7.8 *Top* Space-filling representation of the DNA molecule (Image from the RCSB PDB of PDB ID 5F9I). *Bottom* Ribbon representations of two proteins (Images from the RCSB PDB of PDB ID 3PCY and ID 1AIZ). Created based on images from an *American Scientist* article (Berman et al. 2002, p. 351) (color figure online)

molecule that extend toward the reader in the foreground and to shade those parts that "spiral" into the background. These treatments make the double helical structure of the molecule easier to detect and imagine.

Given these perceptual advantages, it is tempting to conclude that computer-generated 3D models, representing the contemporary technological advancements, should be the visualization of choice in popular science communication for molecular structures. Indeed, computerized molecular modeling is generally hailed as a revolutionary change in the research and education of biochemistry and molecular biology (Pennisi 1994). On the research front, it makes possible the computation of huge amounts of structural data, provides more convenient and versatile visual displays, and allows researchers to predict molecular structures (Pennisi 1994; Tate 2003). On the education front, studies suggest that computer modeling is effective at helping students to envision and understand key concepts, including the complex shapes and structures of molecules and the interactions between molecular compounds (Wu et al. 2001; Dori and Barak 2001; Richardson and Richardson 2002).

At the same time, a higher visual resolution and perceived concreteness do not necessarily lead to public accessibility. 3D molecular models may have dispensed with some of the formal visual symbols used in structural formulas, but that does not mean these models are not created based on codified systems and formal rules. The ribbon model shown in Fig. 7.8 is a case in point. Often seen in popular communication, this model follows specific visualization principles: Important structural elements known as the alpha helices are depicted as ribbons, those known as beta strands are depicted as arrows, and non-repetitive loops are shown as ropes, while other structural details are often omitted (Richardson 1985). Without knowing these conventions (or knowing what alpha helices and beta strands are), a viewer may not gain much, or as much, structural information from these images. Even the space-filling model shown in Fig. 7.8, though seemingly commonsensical in representing atoms as spheres, invokes lesser-known conventions: For example, the spheres' radii are proportional to the radii of the atoms represented, and the center-to-center distance between any two atoms is proportional to the bond strength between the atoms (Bottomley and Helmerhorst 2009; Hardinger 2015). As may be imagined, explanations of such conventions are rarely, if ever, offered in popular communication to accompany 3D models.

In addition, it is important to note that the benefits of 3D molecular models as reported in the contexts of research and formal education do not necessarily transfer to popular communication. An obvious reason, as mentioned above, is that public readers may not know and will not be taught the visual conventions used in molecular modeling. Equally or more importantly, in research and formal education contexts, a viewer has direct access to visualization programs and can directly interact with a model to display it in different styles for select effects, adjust angles and orientations to see a molecule from different perspectives, or highlight and magnify select structural features. In empirical studies that demonstrate the benefits of computer-generated molecular visualization, the benefits come precisely from users being able to view, interact with, and manipulate the models (see Dori and Barak 2001; Wu et al. 2001; Craig et al. 2013). None of these is possible when the models are presented in a magazine or on a static webpage for popular consumption. In these cases, a reader has to perceive depth and spatial arrangements via a flat image to infer unfamiliar 3D structure, which is not an easy task (Roberts et al. 2005).

The rise of computer-generated 3D molecular models in popular communication, then, goes beyond—or cannot be solely attributed to—an attempt for public accessibility. As Kemp (2000) put it, these images are part of an array of "sexy visuals" used to catch the eyes of a large reader base (p. 174). Even in professional publications, such images often adorn journal covers to help project an ethos of technological sophistication and appeal. Indeed, aesthetic consideration seems to have become part and parcel of creating molecular models. In Pennisi's (1994) words, computerized molecular visualization has turned researchers into "artists, sketching with a keyboard instead of pastels. Each new line of computer code, each faster chip has brought a little more of the world beyond our vision into view" (p. 146). Scientists themselves are not unaware of this changed expectation. In Bottomley and Helmerhorst's (2009) comparison of different molecular modeling styles, "visual impressiveness" was cited as a strength for some (e.g., the ribbon style) and the lack of it recognized as a disadvantage for others (e.g., the stick style). As the two authors explained, "an austere, precisely defined, molecular representation would be appropriate for a knowledgeable scientific audience, but a more colorful, and artistic, representation would be appropriate for a nontechnical audience" (p. 261). Not coincidentally, the concept of "molecular art" has become popular in recent years, used by some scientists and artists to refer to stylized renditions of molecular objects with a distinct aim for visual appeal. David Goodsell, structural biologist and artist, is a prominent figure in this effort (see Goodsell 1996, 2010). Examples of molecular art have also been showcased at high-profile events such as the Molecular Graphics Art Show and the IEEE Visualization Conferences (Bottomley and Helmerhorst 2009).

As in the case of micrograph art (see Chap. 3), molecular art is well positioned to brige the age-old gap between arts/humanities and sciences, to spark public attention to and emotional interest in genetics, and to offer opportunities for socially and culturally sensitive visual representations. At the same time, if motivated as strategies to promote popular science publication and the scientific enterprise in general, they risk becoming technologically sophisticated, visually beautiful, yet conceptually vague mysteries. Figure 7.9 is a case in point. Used in a *Science News* article (Barry 2008), Fig. 7.9 employs the ribbon style to show the prospect of using synthetic cells to create ribosomes, the cell unit

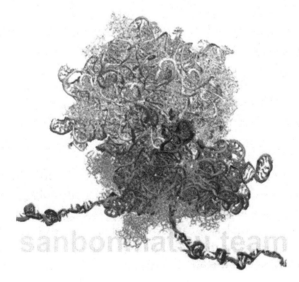

Fig. 7.9 Impressive but vague 3D molecular model (Sanbonmatsu Team, n.d.). © Copyright 2011 Los Alamos National Security, LLC All rights reserved (color figure online)

responsible for turning genetic information into proteins. The image is full of interwoven lines, shapes, and colors and assumes an unusual composition. Awe-inspiring, it is a visual stand-in for the tremendous advancement made in contemporary genetic research. But for readers who are unfamiliar with the modeling mechanism and the molecular substance behind it, it is puzzling what the image actually shows and why it matters. The model, in other words, mystifies more than it reveals. If the change from codified structural formulas to concrete 3D models first strikes one as a move that should empower non-expert readers, images like Fig. 7.9 help to keep that assumption in check.

By this critique, I am not suggesting that computer-generated molecular models are not valuable forms of popular communication. They are. Not only can they create visual interest for reader attention, they afford an immediate and concrete way to "see" invisible molecules, both of which contribute to public engagement. What I hope to do is call attention to 3D models' largely assumed benefits for publics and, more importantly, to the lack of research on their use in popular communication. Current literature on molecular visualization is grounded in the needs of formal research and education. When public readers are occasionally mentioned, scholars (often from computer science and life science) assume that molecular visualization programs, by virtue of their visual and technical advancements, will "engage the public and thereby increase public awareness and understanding of scientific problems" (Hirst et al. 2014, p. 15). But such awareness and understanding, as I argued, will not automatically happen. If current literature has any concrete advice regarding molecular modeling for public readers, it is that these readers be provided with colorful and artistic models without being confronted with technical details. Such advice reflects a rather simplistic understanding of audience that is based purely on perceived "levels" of scientific expertise (see Miller 2004) and harkens back to outdated models of popular science communication that focus on appreciation rather than engagement.

Acknowledging this lack of research, we can more consciously consider how to use molecular visualization in popular science communication, both in genetics and in other subject matters, and whether such use requires local changes or paradigm-level redesign. Based on this study, several heuristics seem warranted.

In terms of presenting 3D models in non-interactive 2D media, explanations, even brief ones, of a model and its noteworthy structure will be useful for readers who lack prior knowledge. For example, with the structure of a binding protein, explanations that what is displayed is a "symmetrical, saddle-shaped" protein that can "ride on" and "bend" DNA (Tjian 1995, p. 60) help readers to selectively process otherwise complex 3D images. In addition, explanations of how a depicted structure impacts genetic function help to reveal the relevance of the model beyond its visual impact. In the same example as quoted above, to say that the bending of the protein and DNA may put gene products in close proximity and in turn facilitate gene transcription shows the significance of the protein structure. Last, verbal descriptions of modeling mechanisms, such as those given earlier about ribbon and space-filling styles, provide some spatial clues for readers to imagine dimensional structures.

These strategies are relatively easy to implement, but they do not solve publics' lack of access to modeling programs and, as such, do not reveal the full benefits of molecular visualization. It is to this larger issue, which I believe significant not only for molecular visualization but public science communication in general, that I now turn.

Web-Based Applets

As mentioned in Chap. 1, current literature on scientific visual representations generally overlooks the needs of public readers. This is especially the case in the area of computer visualization, which was established in the late 1980s and commonly known as visualization in scientific computing or simply ViSC (McCormick et al. 1987). This line of research sees scientific visualization as a "computer-based field" (McCormick et al. 1987, vii), "a method of computing that gives visual form to complex data" (Defanti and Brown 1991, p. 253). Thusly conceived, ViSC attaches quintessential importance to the development of tools—hardware, software, networks, algorithms, and so on. This in turn brings the needs and interests of two stakeholder groups to the foreground: (1) tool makers, "the visualization researchers who [with funding and support] can develop the necessary hardware, software and systems"; (2) tool users, the "experts from engineering and the discipline sciences" who need "better" tools, tools that can, for example, handle massive amounts of data and afford immediate visual feedback (McCormick, et al. 1987, p. 8). In McCormick et al.'s 99-page foundational report that established the ViSC discipline, public readers received only one line, a suggestion that ViSC research strives to "bridge the gap between science and the general public, making individuals aware of significant discoveries" (A-3).

Unfortunately, 20 years later, this call to action remains to be taken up. In a recent *Nature Method* report (Evanko 2010) that reviewed biological visualization, public readers received a single, perfunctory comment: that visual simulation can aid public readers to access scientific methods, models, and hypotheses (Walter et al. 2010, S30). How this may be accomplished was left behind as the report discussed the functions of specialized visualization tools and ways to enhance them for scientists. Likewise, in the particular pursuit of molecular visualization, ongoing research has focused almost exclusively on the concerns of tool makers and tool users: for example, how to make modeling programs more effective, more efficient, and easier for scientists to use (Bottomley and Helmerhorst 2009); how to push for more rapid, massive, and powerful hardware, software, and networks (Hirst et al. 2014); or how to integrate computer modeling into formal education so as to train today's science students to become tomorrow's scientists (Wu et al. 2001; Dori and Barak 2001; Richardson and Richardson 2002; Dori and Kaberman 2012; Jones 2013).

All of these pursuits are, no doubt, quite valuable, but they do not anticipate or accommodate public readers' needs and interests. Given the inherent difficulty of reading 3D images in 2D media and given public readers' lack of codified knowledge and access to cutting-edge technologies, faster, bigger, and more complex tools are less likely to involve publics than a thoughtful adaptation of the tools we already have. This adaptation does not mean we plan to "properly educate" all readers in the formal technologies and processes of computer-based molecular modeling; rather, it means that we invest in giving publics the platforms and opportunities to interact with computer modeling, with the larger goal of raising awareness and interest, as the McCormick et al. (1987) and *Nature* (Evanko 2010) reports would have it.

Such public adaptations may, on first thought, seem unrealistically difficult with knowledge, experience, and technology barriers, but it is useful to note that knowledge and experience are relevant, negotiable concepts. Discipline scientists such as biologists are themselves non-experts when it comes to computer programs, but they manage to work with the programs even as they continue to find some of them confusing and difficult to use (Bottomley and Helmerhorst 2009; O'Donoghue et al. 2010). And as classroom research indicates, pre-collegiate students without advanced knowledge can benefit from using computerized molecular modeling (Dori and Barak 2001; Wu et al. 2001; Dori and Kaberman 2012).

As for technologies, current development in molecular visualization tools has provided the necessary platform for public adaptation. Conventionally, in order to access a modeling program, users must, as Tate (2003) described, obtain (sometimes at high cost) the program and install it locally on a computer (sometimes only on computers with a certain operating system). Though some modeling programs (e.g., Mage and RasMol) are freeware, their setup and use is not straightforward (Weiner et al. 2000). In addition, these programs require users to already possess or be able to obtain input files that contain computerized data about a molecule's structure before that structure can be visually displayed. For novice users, the step of locating and obtaining input files from online databases and navigating multiple file types may already create confusion (Weiner et al. 2000).

Fortunately, the development of the Internet and server-based software has leveled the playing field: "Technologies such as Java, Active X, JavaScript mean that users can run increasingly complex applications without having to explicitly download them, and rely instead on having the application delivered automatically along with a Web page, all for a single click of a mouse" (Tate 2003, p. 145). Java applets, in particular, are compatible with all major browsers and computer operating systems and can provide simple, low-resolution visualizations that are easy to use and deploy (Tate 2003).

Programs that use such technologies to deliver 3D molecular models already exist; one of the most well known is Jmol, which is an open-source browser applet that supports multiple languages (Jmol 2012; MolviZ.org 2015).[4] With pre-built Jmol tutorials, there is no need for software downloading or file inputting. Users can simply open the tutorial in a browser to examine a molecule and interact with it via mouse control. Although lightweight, these tutorials support a range of interactions: for example, switching between different presentation styles such as ribbons and ball-and-sticks, rotating a model in 360° along any axis to view it in different angles, zooming in and out to examine structural details at various magnifications, or highlighting and hiding select elements for specific emphasis—the kinds of interactions that truly take advantage of the benefits of virtual modeling (see Dori and Barak 2001; Wu et al. 2001; Craig et al. 2013).

Figure 7.10, for instance, shows several screenshots from a Jmol tutorial on amino acids, the building block of protein. The human body uses 20 common amino acids, linked in different ways, to produce the staggering number of proteins it needs for proper functioning. The top row

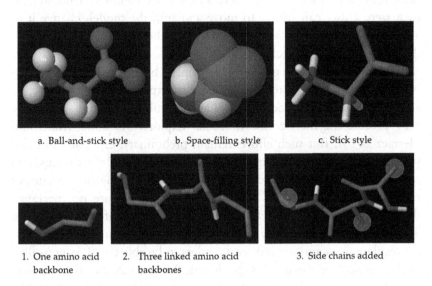

a. Ball-and-stick style   b. Space-filling style   c. Stick style

1. One amino acid   2. Three linked amino acid   3. Side chains added
    backbone   backbones

Fig. 7.10 Molecular modeling as shown in Jmol tutorial (Martz 2007). Courtesy of MolviZ.Org (color figure online)

of Fig. 7.10 shows one of these amino acids, glycine, in three styles: ball-and-stick, space-filling, and stick. As the tutorial explained, these styles have respective limitations and advantages: The ball-and-stick style, for example, reveals the molecule's structure but does not show its relative size and shape; the space-filling style shows size and shape but hides the structural arrangements. These are the reason why comparing multiple styles of the same molecule is deemed essential for viewers to envision 3D structures (Wu and Shah 2004). But such opportunities, as this study showed, are virtually non-existent in print or static online popular science communication, likely because they would require excessive space. With interactive applets, space is not an issue, since at any given time, only one image is displayed on the screen, depending on user choice.

The bottom three images in Fig. 7.10 then demonstrate, in stick style, how individual amino acids can be strung together to form proteins: First displayed is a single amino acid backbone, which is then connected with two identical backbones, and last, side chains are added to the backbone structure to realize a "true" amino acid string. In subsequent steps not shown here, different and more side chain and backbone units are added to form increasingly more complex structures. At each step, a user can "drag" to move and spin the model, change it to a different presentation style, adjust color schemes, hide/reveal atoms, display/remove atom labels, etc.[5]

Currently, tutorials have been developed, based on Jmol as well as other programs, that allow novice users to explore common molecules such as DNA and antibodies as well as molecules of interest such as the HIV protein enzyme (see BioMolecular Explorer 3D 2006; Martz and Herráez 2015). But such efforts remain preliminary, especially in comparison with the development of various structural tools, platforms, and repositories for specialist audiences. In addition, some of the current tutorials, though ostensibly developed for outreach, are not necessarily based on careful consideration of public readers' information needs. The Research Collaboratory tutorials (2016), for instance, forgo narrative explanations and instead trust readers to be able to use jargon-filled menus to explore the models. Such a decision, as studies have shown, may deter readers who are unfamiliar with visualization terminologies (Wu and Shah 2004).

For molecular modeling applets to become a truly engaging tool for public readers, then, we need not only efforts to develop more tutorials, but also research to guide and assess those efforts. Such research has to depart from the current tool-centered concern of ViSC and draw upon theories that allow human-centered studies. Cognitive psychology and information design, given their focus on human cognitive and visual processing, provide promising theoretical frameworks (Supasorn et al. 2008; Jones 2013). The dual-channel cognitive theory, for example, can guide tutorial designers to consider whether and how to juxtapose textual and visual elements. According to this theory, humans process information via the independent yet interactive verbal channel and visual channel (Paivio 1986; Mayer 2009). Delivering information through both channels thus allows a viewer to increase the amount of information gained as well as to integrate verbal and pictorial elements for information processing (Mayer 2009). This is especially relevant when trying to engage novice users who lack prior knowledge in either channel—as in the case of developing modeling applets for publics. At the same time, more information is not necessarily superior. The theory assumes that both channels have a limited capacity and that overloading either channel reduces information processing (Mayer 2009). Acknowledging this constraint can help designers avoid overly ambitious tutorials that, like those developed by FirstGlance (2015) and Research Collaboratory (2016), include numerous menu items and large amounts of background information.

The dual-channel theory also has implications for *how* verbal and visual information may be correlated in modeling applets. As Mayer and Moreno (2003) wrote, when complex verbal and visual information is presented in user-controlled segments, it allows users to process information one step at a time and avoid cognitive overload. At each step, verbal and visual information is ideally synchronized so viewers do not need to "hold" one piece of information in their working memories until the other is given (Mayer and Moreno 2003). According to these understandings, tutorials that break down the process of how complex molecules are formed and explain each step with synchronized verbal and visual information—as in Fig. 7.10 (also see Martz and Herráez [2015])—are more likely to increase public access, participation, and eventual engagement.

Beyond verbal and visual correlation, various other factors need to be examined. For example, would readers prefer any particular presentation styles? Research conducted in formal education contexts suggests that students favor the ball-and-stick style and find the space-filling style less illuminating even though more attractive (Wu et al. 2001; Dori and Kaberman 2012). Future research may examine whether the same holds true in the context of public science communication and why. In addition, it is useful to examine whether readers would prefer viewing molecular structures in particular research/application contexts (e.g., how a drug's structure allows it to inhibit disease progression) or would they be interested in exploring elements fundamental to genetics (e.g., DNA and RNA)? Findings from these and other related inquires will have important implications for the future development of public-facing, interactive molecular modeling applets.

Conclusion

To be able to examine and manipulate the structures of complex molecular compounds is a cornerstone of biochemical research. In the narrower field of genetics, structural modeling techniques afforded one of the most publicized breakthroughs of all times, facilitated modern genetic research, and informed contemporary genome-wide studies. As stakeholders ranging from students, educators, and scientists in life and natural science disciplines benefit from these advancements, it is also the time that we consider the role these techniques can play in public science communication.

Such considerations, as this chapter shows, quickly reveal the formally codified nature of structural images and, along with that, a tacit requirement for prior knowledge from viewers. This reality has not changed as visualization techniques and formats shifted from pen-and-paper to computers, from 2D schematics to 3D high-definition models. In fact, one may say that the development of computer modeling has widened the gap between specialists' and publics' technological access to structural information. For these reasons, structural images as they have developed so far will not be automatically accessible or relevant to public readers—except in the limited sense of creating visual appeal. If we want publics to appreciate the relevance of structural information, to recognize the importance of structural examination, and to actually engage in such efforts, molecular visualizations must draw upon readers' innate visual experience and kinetic skills. With 2D structural formulas, this means designing the images not so much as codified lines and shapes but as meaningful visual patterns; with 3D models, this means presenting molecules not (or not only) as attractive visual feasts but interactive platforms that afford direct interaction. For the newer generations of readers who grew up with the Internet and server-based applications, such a platform can be especially familiar and welcomed.

Of course, these are far from the only ways to approach molecular structures in popular communication, and that is precisely what this chapter argues: current research is sorely lacking and existing work either ignores or gives lip service to the supposed usefulness and benefits of molecular visualization to public readers. To begin to change this reality, we need concerted efforts in visual design, technology development, and targeted research. Given its nature, such research will inevitably be multidisciplinary and draw upon theories in public science communication, scientific visual communication, cognitive psychology, and information design, among others. To be impactful, it also needs to be eclectic in its questions and approaches and rely on rhetorical analyses, empirical studies, quantitative measurements, qualitative inquiries, reviews, and meta-analyses, among others.

Notes

1. The octet rule is not applicable to all situations and certain exceptions exist (see Chang 2003).
2. Certainly, another important factor is the rise of 3D modeling, a topic discussed later in the chapter.
3. In addition to omitting carbons and carbon–hydrogen bonds, Fig. 7.3 also omitted the chemical details of the DNA bases, as those are not important for conveying its point.
4. Jmol also provides a separate, downloadable application.
5. The same is true with the top three displays in Fig. 7.10.

References

Amare, N., & Manning, A. (2013). *A unified theory of information design: Visuals, text & ethics*. Amityville, NY: Baywood Publishing.

Baake, K. (2003). *Metaphor and knowledge: The challenges of writing science*. Albany, NY: State University of New York Press.

Barry, P. (2008). Life from scratch: Learning to make synthetic cells. *Science News, 173*(2), 27–29.

Bearn, A. G. (1956). The chemistry of hereditary disease. *Scientific American*, 126–136.

Berman, H. M., Goodsell, D. S., & Bourne, P. E. (2002). Protein structures: From famine to feast. *American Scientist, 90*(4), 350–359.

BioMolecular Explorer 3D. (2006). Retrieved February 22, 2016, from http://www.umass.edu/molvis/bme3d/materials/explore.html.

Bottomley, S., & Helmerhorst, E. (2009). Molecular visualisation. In J. Gu & P. E. Bourne (Eds.), *Structural bioinformatics* (2nd ed., pp. 237–268). Hoboken, NJ: Wiley-Blackwell.

Bradshaw, J. (2001). Strings and things: A brief history of chemical languages. Retrieved February 20, 2016, from http://www.daylight.com/meetings/mug01/Bradshaw/History/.

Calvin, M., & Calvi, G. J. (1964). Atom to Adam. *American Scientist, 52*(2), 163–186.

Chang, R. (2003). *Chemistry* (8th ed.). New York: McGraw-Hill.

Christiansen, J. (2013). A defense of artistic license in illustrating scientific concepts for a non-specialist audience. In *COmmunicating COmplexity 2013 Conference Proceedings* (pp. 49–60). Rome: Edizioni Nuova Cultura-Roma.

Cook, M. P. (2006). Visual representations in science education: The influence of prior knowledge and cognitive load theory on instructional design principles. *Science Education, 90*(6), 1073–1091. doi:10.1002/sce.20164.

Cooper, M. M., Grove, N., Underwood, S. M., & Klymkowsky, M. W. (2010). Lost in Lewis structures: An investigation of student difficulties in developing representational competence. *Journal of Chemical Education, 87*(8), 869–874.

Craig, P. A., Michel, L. V., & Bateman, R. C. (2013). A survey of educational uses of molecular visualization freeware. *Biochemistry and Molecular Biology Education, 41*(3), 193–205. doi:10.1002/bmb.20693.

DeFanti, T. A., & Brown, M. D. (1991). Visualization in scientific computing. *Advances in Computers, 33*(1), 247–305.

Dori, Y. J., & Barak, M. (2001). Virtual and physical molecular modeling: Fostering model perception and spatial understanding. *Educational Technology & Society, 4*(1), 61–74.

Dori, Y., & Kaberman, Z. (2012). Assessing high school chemistry students' modeling sub-skills in a computerized molecular modeling learning environment. *Instructional Science, 40*(1), 69–91. doi:10.1007/s11251-011-9172-7.

Double Helix at the Atomic Level. (1973). *Science News, 103*(17), 267–268.

Evanko, D. (2010). Supplement on visualizing biological data. *Nature Methods, 7*(3), S1–S68 and figures.

FirstGlance in Jmol. (2015). Retrieved June 3, 2016, from http://bioinformatics.org/firstglance/fgij/gallery.htm?NOJAVA.

Friedmann, T. (1971). Prenatal diagnosis of genetic disease. *Scientific American,* 34–42.

Gallo, R. C. (1986). The AIDS virus. *Scientific American,* 46–56.

Gasgall, M. (2008). *Ribbon diagrams.* Retrieved June 18, 2016, from https://research.duke.edu/ribbon-diagrams.

Goodsell, D. S. (1996). *Our molecular nature: The body's motors, machines and messages.* Göttingen, Germany: Copernicus.

Goodsell, D. S. (2010). *The machinery of life.* Göttingen, Germany: Copernicus.

Gross, A. G. (2007). Medical tables, graphics and photographs: How they work. *Journal of Technical Writing and Communication, 37*(4), 419–433.

Hardinger, S. (2015). *Illustrated glossary of organic chemistry.* Retrieved July 19, 2017, from http://www.chem.ucla.edu/harding/.

Hershey, A. D., & Chase, M. (1952). Independent functions of viral protein and nucleic acid in growth of bacteriophage. *The Journal of General Physiology, 36*(1), 39–56.

Hirst, J. D., Glowacki, D. R., & Baaden, M. (2014). Molecular simulations and visualization: Introduction and overview. *Faraday Discussions, 169,* 9–22. doi:10.1039/c4fd90024c.

Image from the RCSB PDB (http://www.rcsb.org) of PDB ID 3PCY. (Church, W. B., Guss, J. M., Potter, J. J., & Freeman, H. C. (1986). The crystal structure of mercury-substituted poplar plastocyanin at 1.9-A resolution. *The Journal of Biological Chemistry, 261,* 234–237).

Image from the RCSB PDB (http://www.rcsb.org) of PDB ID 5F9I. (Garcia, S., Acosta-Reyes, F. J., Saperas, N., & Campos, J. L. (2015). Crystal structure of rich-AT DNA 20mer. To be published).

Image from the RCSB PDB (http://www.rcsb.org) of PDB ID 1AIZ. (Shepard, W. E., Kingston, R. L., Anderson, B. F., & Baker, E. N. (1993). Structure of apo-azurin from Alcaligenes denitrificans at 1.8 A resolution. *Acta Crystallographica Section D, 49,* 331–343).

Jmol. (2012). An open-source Java viewer for chemical structures in 3D. Retrieved July 25, 2016, from http://jmol.sourceforge.net/.

Johnson, J. (2010). *Designing with the mind in mind: Simple guide to understanding user interface design rules.* Burlington, MA: Morgan Kaufmann.

Jones, L. L. (2013). How multimedia-based learning and molecular visualization change the landscape of chemical education research. *Journal of Chemical Education, 90*(12), 1571–1576.

Kekulé, F. A. (1872). Ueber einige Condensationsproducte des Aldehyds. *Justus Liebigs Annalen der Chemie, 162*(1), 77–124.

Kemp, M. (2000). *Visualizations: The nature book of art and science.* Oakland: University of California Press.

Lewis, G. N. (1916). The atom and the molecule. *Journal of the American Chemical Society, 38*(4), 762–785.

Martz, E. (2007). Protein structure. Retrieved August 1, 2016, from http://www.umass.edu/molvis/bme3d/materials/jmoltuts/antibody/contents/contents.htm.

Martz, E., & Herráez, A. (2015). DNA structure tutorial. Retrieved August 4, 2016, from http://biomodel.uah.es/en/model4/dna/index.htm.

Mayer, R. (2009). *Multimedia learning* (2nd ed.). New York: Cambridge University Press.

Mayer, R. E., & Moreno, R. (2003). Nine ways to reduce cognitive load in multimedia learning. *Educational Psychologist, 38*(1), 43–52. doi:10.1207/S15326985EP3801_6.

McCormick, B. H., DeFanti, T. A., & Brown, M. D. (1987). Visualization in scientific computing. *Computer Graphics, 21*(6), i-E8.

Miller, J. A. (1977). Cancer clues from chemical structures. *Science News, 111*(23), 362–363.

Miller, G. A. (1994). The magical number seven, plus or minus two: Some limits on our capacity for processing information. *Psychological Review, 101*(2), 343–352.

Miller, C. (2004). A humanistic rationale for technical writing. In J. Johnson-Eilola & S. A. Selber (Eds.), *Central works in technical communication* (pp. 47–54). New York: Oxford University Press.

MolviZ.org: Molecular Visualization Resources. (2015). Retrieved August 3, 2016, from http://www.umass.edu/microbio/chime/index.html.

O'Donoghue, S. I., Goodsell, D. S., Frangakis, A. S., Jossinet, F., Laskowski, R. A., Nilges, M., ... Olson, A. J. (2010). Visualization of macromolecular structures. *Nature Methods, 7*(3), S42–S55.

Paivio, A. (1986). *Mental representations*. New York: Oxford University Press.

Peirce, C. S. (1894). What is a sign? Retrieved June 18, 2015, from http://www.iupui.edu/~arisbe/menu/library/bycsp/bycsp.HTM.

Pennisi, E. (1994). Twirling ribbons, billowing bubbles. *Science News, 146*(21), 328–330.

Pray, L. A. (2008). Discovery of DNA structure and function: Watson and Crick. *Nature Education, 1*(1), 100.

Research Collaboratory for Structural Bioinformatics (RCSB) Protein Data Bank (PDB)-101. (2016). Retrieved August 10, 2016, from http://www.rcsb.org/pdb/secondary.do?p=v2/secondary/visualize.jsp#visualize_jmol.

Richardson, J. S. (1985). Schematic drawings of protein structures. *Methods in Enzymology, 115,* 359–380.

Richardson, D. C., & Richardson, J. S. (2002). Teaching molecular 3-D literacy. *Biochemistry and Molecular Biology Education, 30*(1), 21–26. doi:10.1002/bmb.2002.494030010005.

Roberts, J. R., Hagedorn, E., Dillenburg, P., Patrick, M., & Herman, T. (2005). Physical models enhance molecular three-dimensional literacy in an introductory biochemistry course. *Biochemistry and Molecular Biology Education, 33*(2), 105–110. doi:10.1002/bmb.2005.494033022426.

Sanbonmatsu Team. (n.d.). Log Alamos National Laboratory. Images. Retrieved July 29, 2016, from http://sanbonmatsu.org/image2.html.

Schlecht, M. F. (1997). *Molecular modeling on the PC*. Weinheim, Germany: Wiley-VCH.

Supasorn, S., Suits, J. P., Jones, L. L., & Vibuljan, S. (2008). Impact of a pre-laboratory organic-extraction simulation on comprehension and attitudes of undergraduate chemistry students. *Chemistry Education Research and Practice, 9*(2), 169–181. doi:10.1039/b806234j.

Sweller, J. (1994). Cognitive load theory, learning difficulty, and instructional design. *Learning and Instruction, 4*(4), 295–312.

Tate, J. (2003). Molecular visualization. *Methods of Biochemical Analysis, 44,* 135–158.

Tjian, R. (1995, February). Molecular machines that control genes. *Scientific American,* 54–61.

Walter, T., David, W. S., Baldock, R., Mark, E. B., Anne, E. C., Duce, S., ... Hériché, J. (2010). Visualization of image data from cells to organisms. *Nature Methods, 7*(3), S26–S55. doi:10.1038/nmeth.1431.

Ware, C. (2012). *Information visualization: Perception for design* (3rd ed.). Burlington, MA: Morgan Kaufmann.

Weiner, S. W., Cerpovicz, P. F., Dixon, D. W., Harden, D. B., Hobbs, D. S., & Gosnell, D. L. (2000). RasMol and Mage in the undergraduate biochemistry curriculum. *Journal of Chemical Education, 77*(3), 401–406.

Wu, H., & Shah, P. (2004). Exploring visuospatial thinking in chemistry learning. *Science Education, 88*(3), 465–492. doi:10.1002/sce.10126.

Wu, H., Krajcik, J. S., & Soloway, E. (2001). Promoting understanding of chemical representations: Students' use of a visualization tool in the classroom. *Journal of Research in Science Teaching, 38*(7), 821–842. doi:10.1002/tea.1033.

8

Conclusion

Freeman Dyson, renowned physicist and mathematician, declared in his 2015 book *Dreams of Earth and Sky* that "the domestication of biotechnology will dominate our lives during the next fifty years at least as much as the domestication of computers has dominated our lives during the previous fifty years" (p. 1). With this forecast, Dyson predicted that in the near future, there will be

> do-it-yourself kits for gardeners who will use genetic engineering to breed new varieties of roses and orchids. Also kits for lovers of pigeons and parrots and lizards and snakes to breed new varieties of pets.... Domesticated biotechnology, once it gets into the hands of housewives and children, will give us an explosion of diversity of new living creatures.... New lineages will proliferate to replace those that monoculture farming and deforestation have destroyed. Designing genomes will be a personal thing, a new art form as creative as painting or sculpture. (pp. 2–3)

Dyson's claims are bold and fraught with ethical and social ramifications, which, by the way, he wasn't ready to address. But the prediction of a future where biotechnology infuses every aspect of our life is not groundless. In the USA, the majorities of some major crops (including

© The Author(s) 2017
H. Yu, *Communicating Genetics*,
DOI 10.1057/978-1-137-58779-4_8

corn, cotton, and soybean) are already genetically modified. In the next 25 years, genetically modified forest trees will move from research to application (Häggman et. al 2016). In the animal world, a variety of transgenic species have been produced since the first of them, transgenic mice, was born in 1974. And in late 2015, the U.S. Food and Drug Administration approved the first genetically designed animal for public consumption: the fast-growing AquAdvantage salmon. Although we do not usually speak of genetically modified humans, that is precisely what genetic therapies are about: repairing, editing, deleting, or inserting genes in the human genome in the hope of preventing and treating diseases. Such experimentation is accompanied and bolstered by the fast growth in personalized genetic/genomic testing and initiatives such as precision medicine. Today, for only $99, consumers can obtain DNA tests with more than 700,000 genetic markers to explore their ethnic background and family history (AncestryDNA 2016). For $999 (and a doctor's order), they can obtain a full-genome sequencing report to explore health risks and develop personalized medical and lifestyle protocols (Veritas Genetics 2016). Last, and probably most tellingly with regard to Dyson's prediction, a growing DIYbio community is enthusiastically pursuing synthetic biology. Comprised of both professionals and non-experts, these DIYers use self-made tools and household items (as well as professional equipment and materials) to carry out procedures ranging from DNA extraction and purification to genetic testing (Kuznetsov et. al 2012).

With all of these signs, I am inclined to believe Dyson that the next few decades will see tremendous integration of biotechnology in average citizens' everyday lives, if not in the kind of recreational use Dyson imagined, then certainly in the realms of personal identity, health care, and lifestyle choices. With such a future, genetics will be a concrete and real part of our life, not an esoteric discipline being worked away in the laboratory. With such a future, we cannot afford *not* to seriously consider today and tomorrow's public communication of genetics.

This communication, whether in verbal, visual, or other multimedia forms, needs to convey pertinent and non-simplistic information. It needs to make everyday citizens, not just "elite" audiences or "fans" of science, feel adequately informed and empowered to make personal decisions about genetically modified organisms, genetic testing, and genetic medicine. Ideally, it will also encourage and inspire citizens to broach public discussions and debates about genetic research and the ethical and social consequences of biotechnology. But simply delivering knowledge so publics have cognitive understanding of terminologies or research protocols is not enough. Such a narrow agenda, as social scientists have shown, is problematic for the communication of any scientific information. It is especially inadequate with regard to genetics, which has inherently affective and social–cultural dimensions. After all, this is a discipline that studies lives—including the most precious life to us, our own—as well as nature, to which we have not only physical but spiritual attachments and ethical responsibilities. It is a discipline that deals with life's risks and promises, which are inevitably emotional and contingent, and with individuals and identities, which are social–cultural constructs as much as they are biological ones.

With all of these layers and factors, visual representations of genetics stand to offer a compelling area for investigation, because images are direct embodiments of the cognitive, the affective, and the social-cultural dimensions. They are cognitive because vision provides us with more information than all of the other senses combined and is a fundamental part of our cognitive activity (Ware 2012). They are affective because they approximate, at least in syntactic choices, the external world we experience and thus elicit the emotions we project onto that world. They are social–cultural because, as semiotic signs, they are motivated by sign-makers' specific interests, or the lack thereof, and can evoke the same among otherwise unmotivated viewers. When one considers these functions of images, it is surprising that more studies have not been devoted to the visual communication of genetics in the public domain.

The lack of this research is partly because of tradition: rhetoricians, science studies scholars, and communication scholars are traditionally trained in verbal discourses, are comfortable with and passionate about language-based studies, and have established recognizable and accepted research methods, theories, and premises (which, by the way, have made the undertaking of this project a risk on my part).

Another reason for the lack of visual-oriented studies is the misconception shared by some that images are easier to process than obscure terminologies and impenetrable prose; after all, one only needs to look at them to "take it all in." But this, as I show in this book, is far from the case. In modern science, seeing is complicated by the use of instruments: electron microscopes, DNA sequencers, or simply statistical analyses, which transform invisible data, objects, and phenomena into images that are fundamentally different from those seen in our macroscopic world. What we end up seeing is therefore not an object per se but "an interaction between an instrument and the object to which it is applied, or a set of parameters that describe such an interaction" (Pomian 1998, p. 227). To borrow Latour's (1998) term, what we see is a black box that has eclipsed the object being studied, the motivation behind the study, and the human interest and subjectivity associated with the study.

These complications and hidden elements are what this book attempts to reveal. It is by doing so, by demonstrating that we cannot take images for granted, that we may hope to see more research off the conventional path, more research that, in their own ways, unpack popular science visual communication. When carrying out such studies, we have to accept, as the previous chapters show (probably to some readers' disappointment and certainly to my own revelation), that there are often few hard and fast "rules" regarding the design of popular science images. Few rules exist because cognitive, affective, and social–cultural dimensions interact with each other and point to varying design directions. For example, while the cognitive function of an image may be more easily fulfilled by reducing data complexity, such a decision may not be socially responsible in terms of informing and empowering the publics. Or, the affective dimension of an image may be easily realized via aesthetic choices, but such a decision may overshadow cognitive relevance and, indeed, may be so successful at promoting science that it overshadows the social implications of science.

Even within a single dimension, multiple factors preclude specific "best practices." In designing for cognitive accessibility, for example, one needs to consider whether a visual genre is familiar or unfamiliar to some and all publics, whether the information in question is abstract or concrete, and how much information needs to be presented for the image to be cognitively interesting without being overwhelming. With regard to affect, there are at least three interconnected but different levels to consider: There are the emotions evoked by the portrayed/perceived humans, animals, and other agents involved in genetic research; there are the emotions infused by communicators and scientists to attract public readers; and there are the emotions publics bring with them to viewing the images. Last, with social–cultural considerations, one must wrestle with the fact that it is not always easy or even possible to determine what is and is not socially responsible and culturally sensitive: For example, can images be used, based on select evidence, to encourage public trust, admiration, and appreciation in/of science? What about doing the same to encourage public mistrust, critique, and monitoring? Where do images cross the line between being authoritative and deterministic, between being democratic and patronizing?

To address these and other such questions and complexities, the book has relied on detailed and contextualized examples and multi-perspective analyses. Without being essentializing, the book as a whole also points to some higher-level observations and suggestions. First, it demonstrates that as genetic research advanced and especially as it transitioned from classical to molecular research, visuals necessarily took on more cognitive baggage and disciplinary convention. When this happened, trying to convey the full range of research to non-expert readers became increasingly challenging (such as with structural formulas and molecular models) and requires careful balancing between information richness and publics' information needs. On this front, post-1980s visual representations, compared with earlier works, are more self-conscious and successful (as in the cases of adapted Cartesian graphs and narrative-based micrographs). At the same time, not all efforts are equally successful (consider the simplistic representations of the genetic code), and still others are missed opportunities (such as pictographs and pattern-based structural formulas). As I argued in earlier chapters, packaging abstract, unfamiliar, and invisible scientific information in accessible and relatable ways for non-expert audiences is the very essence of contemporary public science communication. Ongoing research, in regard to both media production and public reception, is needed to negotiate this challenge.

Second, the book demonstrates that contemporary visual creators are more conscious in expressing affective appeals and relating to readers' emotional interest. Noticeably, through the use of rich syntactic details and imageries, today's visual representations are more affectively relevant—whether that means aesthetically beautiful or intriguing or familiar or welcoming. These changes are to be applauded, as they speak to the inherent connections between sciences, humanities, and arts and allow science communication to compete with other mass media content for reader attention. At the same time, these changes are not without possible risks. When rich colors, bold lines, spectacular compositions, and metaphorical imageries become routine practices, one cannot help but wonder whether we are turning public science communication into more of a spectacle than anything else—and if so, with what consequences. Would it reduce visual evidence (as with decontextualized micrographs), discourage active inquiries (as in the case of metaphorical illustrations), or promote a deterministic and reductive attitude toward genetics (as with symbolic photographs)? More research is needed to examine, contemplate, and debate what the "proper" balance is between affect and cognition, whether such a balance exists (if not in general, then in situated contexts), and if so, how to realize that balance.

Last, the book argues that digital visualization is an area that deserves immediate attention. Database visualization tools, games, and Web applets ushered in new and exciting prospects for the visual representation of genetics: they enhance data sharing and access, lower the threshold for hardware and software, are superior at handling big data, and allow user interaction and customized visualization. It is, however, equally disappointing to see that while tremendous efforts and attention went into developing and refining digital tools for scientists, specialists, and formal education, very little went into serving the publics. Indeed, there is a complacent belief (or denial) in current literature that tools with high resolution, speedy calculation, powerful functions, and Internet-based connectivity will automatically create a superior and egalitarian experience for citizens. As I argued in earlier chapters, this is far from being true. To encourage genuine public engagement, we need more than crowd-sourcing games, dazzling 3D images, and public domain datasets. Significant research and investment are needed to understand publics' unique needs in multimedia tools and to develop those tools for distribution, feedback, and refinement.

Connected to this line of research and investment is a particular area of interest: mobile apps. As of 2015, smartphone subscription reached 64% in the USA, and US-based Internet users were also spending more time online via mobile devices than via desktops and laptops (Meeker 2015, p. 14, 117). These statistics point to the promise of developing mobile apps for popular science communication. Indeed, a quick search in the iTunes app store already returns a decent number of apps that deal with genetics. The National Human Genome Research Institute (2011), for example, publishes a Talking Glossary of Genetic Terms app. The app opens with a long list of alphabetized terms that are supposed to be "commonly used today in news reports, by researchers and medical professionals, in classrooms and, increasingly, as part of daily conversation." Each term, when clicked, leads to a set of audio, text, and/or visual based explanation. While this use of multimodal explanation caters to audiences with different learning styles, none of the explanation allows user interaction, which forfeits an important advantage of mobile computing at engaging users. In particular, the glossary-based interface reflects a naive conviction about (and deficit approach to) public science communication. The developers seem to assume that users will either study the glossary from cover to cover or come to this resource with a given term in mind. Neither assumption seems particularly realistic: Few public users will likely be interested in studying through a list of decontextualized terms, and if they have questions about, say, media reports of genetics, those questions are likely to be much messier than a neatly packaged term. At the very least, quick online searches for terminologies as one encounters them seem far more convenient. In short, as with other digital tools, mobile computing alone does not guarantee user engagement, and critical examination is needed for the future development of this platform.

What mobile computing did do is revolutionize the way we access and share information, the way we work, play, socialize, and live everyday life. This is something we would not have imagined even 10, 15 years ago. If Dyson is right, then the change brought by genetics and biotechnology in the near future will be even more impactful and far-reaching. Even for those who are not inclined to predict the future, ours is already a time when genetic research reshapes the diagnosis and treatment of disease, changes what we put on the dinner table, and complicates our perception of identity and equality. These developments and prospects call for scholars from various disciplines and research paradigms to investigate and ponder the popular communication of genetics, and I hope this book has contributed to that effort.

References

AncestryDNA. (2016). Discover the family story your DNA can tell. Retrieved May 25, 2016, from http://dna.ancestry.com/.

Dyson, F. (2015). *Dreams of earth and sky*. New York: New York Review Book.

Häggman, H., Sutela, S., & Fladung, M. (2016). Genetic engineering contribution to forest tree breeding efforts. In C. Vettori et al. (Eds.), *Biosafety of forest transgenic trees* (pp. 11–29). Netherlands: Springer.

Kuznetsov, S., Taylor, A. S., Regan, T., Villar, N., & Paulos, E. (2012). At the seams: DIYbio and opportunities for HCI. In *Proceedings of the Designing Interactive Systems Conference* (pp. 258–267). New York: ACM. doi:10.1145/2317956.2317997.

Latour, B. (1998). How to be iconophilic in art, science, and religion. In C. A. Jones, P. Galison, & A. E. Slaton (Eds.), *Picturing science, producing art* (pp. 418–440). New York: Routledge.

Meeker, M. (2015, May 27). Internet trends 2015—code conference. Retrieved May 18, 2016, from http://www.kpcb.com/blog/2015-internet-trends.

National Human Genome Research Institute. (2011). Talking glossary of genetic terms (Version 1.0.3) [Mobile Application Software]. Retrieved May 25, 2016, from http://itunes.apple.com.

Pomian, K. (1998). Vision and cognition. In C. Jones & P. Galison (Eds.), *Picturing science, producing art* (pp. 211–231). New York, NY: Routledge.

U.S. Food and Drug Administration. (2015, November 19). FDA has determined that the AquAdvantage salmon is as safe to eat as non-GE salmon. Retrieved May 20, 2016, from http://www.fda.gov/ForConsumers/Consumer Updates/ucm472487.htm.

Veritas Genetics. (2016). Veritas myGenome. Retrieved May 20, 2016, from https://www.veritasgenetics.com/mygenome#sec-4.

Ware, C. (2012). *Information visualization: Perception for design* (3rd ed.). Burlington, MA: Morgan Kaufmann.

Index

© The Editor(s) (if applicable) and The Author(s) 2017
H. Yu, *Communicating Genetics*,
DOI 10.1057/978-1-137-58779-4

Printed in the United States
By Bookmasters